BOYS AND OTHER ANIMALS

Published in 2002 by

WOODFIELD PUBLISHING
Woodfield House, Babsham Lane, Bognor Regis
West Sussex PO21 5EL, England.

ISBN 1-903953-27-8

Boys and Other Animals

JOSEPHINE DUGGAN REES

Woodfield Publishing

1960 – Peter, Nick and Nigel with Beauty's puppies.

1-year-old Jonathan and friends, 1962

CONTENTS

Dug and boys swimming at Climping.

The family choir at St Mary's Slindon.

Foreword

In this, her third book for Woodfield,[1] Josephine Duggan Rees takes a light-hearted look back at her experiences of family life and motherhood in the 1950s-80s.

In 1952 her husband Walter was appointed farm manager at Court Hill Farm in the Sussex village of Slindon and for the ensuing 27 years they lived in a rambling farmhouse in the heart of the English countryside, raising an energetic family of four boys whilst overseeing the running of a large arable and livestock farm.

The frequently comical trials and tribulations of their life on the farm – involving a large cast of colourful rural characters and a menagerie of assorted dogs, cats and horses – are delightfully depicted in a light-hearted style that will be familiar to the many readers who have already enjoyed *Corduroy Days* (Josephine's entertaining account of her experiences as a 'land girl' in the Women's Land Army during World War II).

Josephine's struggle to keep up with the demands made on her as a farmer's wife and mother of a boisterous all-male family is the source of much amusement. Always willing to laugh at herself, she makes frequent fun of her own failings as a cook, housekeeper and childminder whilst describing the many escapades of her growing boys – their antics as toddlers, their 'adoption' of all manner of stray creatures as pets, their hare-brained money-making schemes as teenagers – and much else. All is recounted with a keen eye for observation and a gentle humour which will delight readers of all ages and particularly those with a farming background.

---

[1] See page 247 for details of A *Portrait of Slindon* and *Corduroy Days*.

The Author on Stobie, watched by boys.

Nigel with friend

1. Back to Sussex

We left the flint walls and cottages of the village and the tall Tudor chimneys of the manor house behind, and turned into a steep, winding lane that led to the farm. Beech trees spread a canopy overhead, already tinged with autumn colour, and in a field below a plough carved brown stripes through the golden stubble. Then, as the road levelled and straightened, we passed a farm pond and a cluster of flint buildings. Beyond them stood the farmhouse, rather imposing I thought, with its white Georgian façade, surrounding trees and apron of gravel drive. Fields sloped upward from it on either side to wooded hilltops, while in the distance rolled the Downs, clad in purple mist.

"It's good to be back in Sussex," I said to Dug, remembering the days of my childhood, walking on Beachy Head and Devil's Dyke. There had since been sojourns in many southern countries, including Somerset and Oxfordshire. Now I was home.

"Isn't it lovely?" I asked.

Dug grunted. A Welshman, born in Shropshire, he declared he had found nothing of consequence south of that county, with the possible exception of myself, and he did not share my admiration of the South Downs. True, these here had not the grandeur of those I had known in East Sussex but they were imposing nonetheless.

"Huh, mole hills!" he scoffed. "You should see the Welsh mountains."

I had seen the Welsh mountains, and very grand and beautiful they are, when not obscured by rain and mist. The weather is kinder in the south.

"Anyway, I haven't got the job yet and I may not want it."

The job was managing this near thousand-acre mixed farm with a three thousand acre shoot.

We drove into the yard, parking our modest Hillman a discreet distance from a shiny black Daimler. My husband's prospective employer, whom I will call Mr Long, because that was not his name, got out and walked to meet us; a man of about Dug's age, of medium height, fair haired with ruddy complexion.

He eyed us, I thought, with approval, especially me in my plain, tweed suit and brogue shoes; no doubt I looked a 'sensible' type.

"You can look at the house first," he said, after the preliminaries. "Come back to the yard when you have finished."

A stout, cheerful-looking woman with grey hair swept carelessly into a bun, answered the door.

"Shall we start at the top?" she asked.

The top was reached by two flights of stairs with several twists and turns and consisted of three attic bedrooms, a box room and a landing cupboard.

We were admiring the view from one of the dormer windows while getting our breath when the telephone rang in the hall and our guide excused herself.

"It always rings when I'm up here," she sighed.

"I have to take a message," she called up a moment later. "Won't be a minute."

From the window we saw her hurrying along the road to the buildings. It occurred to me that answering the telephone and taking messages could be a time-consuming part of the manager's wife's day.

We met on the first-floor landing to resume the tour of the bedrooms. Downstairs we found a number of children, cats and kittens in assorted sizes, about five or six of each at a rough count. They took little notice of us.

In the kitchen I had the sensation of taking a trip backwards in a time machine. Judging by what we had previously seen of the house, I had dated it early nineteenth century, but in these back premises the low ceiling, shuttered windows, old flagstones that paved the floor and deep chimney recess that might have once housed a spit when the house was new, were evocative of a much earlier period.

At one end it lead into a brick-floored larder and pantry, at the other into a washhouse and down steps into a large brick floored area where a concrete slab covered a deep well.

It was an austere house with a presence of the past; a happy past where many large families had grown up with little material wealth perhaps, but taking their pleasure from the woods and Downs around them. I wanted to bring up our family here and I was happy when Dug told me that he had been offered the job and decided to take it.

We came to Court Hill Farm a few weeks later, accompanied by our three sons, Nigel (aged six), Nicholas (three-and-a-half) and Peter (two), Mary, my teenage 'mother's help', Jane (our golden retriever, who was pregnant) and six hens. In this coronation year, 1952, which had seen a new queen on the throne, we were about to start a new life…

The beeches were now resplendent in hues of amber and bronze, blackberries still hung lush on crimson briars in the hedgerows, mingled with skeins of the white bryony's orange berries, the scarlet hawthorn and the soft, white canopy of old, man's beard. A smell of autumn hung on the air: of wet leaves, berries and wood-smoke; apples hung thick on the gnarled trees in the orchard.

Once our furniture was in, the house looked welcoming. A fire, kindly lit by the herdsman's wife, was burning cheerfully in the new Rayburn that replaced the old range that I had viewed with some trepidation on our first visit. A can of milk stood on the table.

Once a fire is lit and a meal prepared, however makeshift, a house becomes home, even if most of one's possessions are still in packing cases. I quickly went out and picked a few of the Michaelmas daisies growing in the untidy flowerbeds and put them in an earthenware vase I found in the larder. Now the picture was complete. The next task was choosing bedrooms and making up the beds.

Dug, meanwhile had gone out to meet the men and tour the farm.

"I'm going to sleep at the top," announced Nigel, who had already explored the house.

"By yourself?" I queried.

"Yes. I'm not going to sleep with the babies," he said, with his usual disdain for his younger siblings.

So I made up a bed for my eldest son in one of the front attic bedrooms and in the large room below for Dug and myself, while the small children were to sleep opposite and Mary in the one behind it, which had a communicating door into what became known as 'the nursery'.

We all slept well that first night in the strange house, with the possible exception of Nigel, who, in his chosen isolation, had heard strange rustlings, scratchings and gurgles, which we accounted for by mice in the skirting, birds in the roof and air in the cold water tank in the box room.

A few nights later he declared that he had seem bright yellow eyes glaring at him in the darkness. This I put down to imagination or perhaps a cat that had strayed upstairs, for we had taken over the resident collection, until, the following morning I met a large rat at the top of the stairs.

That night it took a fancy to regency striped fabric and gnawed six inches off the bottom of one of my new dining room curtains. The rats were hastily dealt with and Nigel complained of no more gleaming eyes and scratching in the night. Soon, however, we were hosts to another species of vermin. Every day I found a broken egg in the nest boxes and suspecting a miscreant among the hens I watched them carefully. I discovered that my egg eater had a long busy tail.

Every day after that we saw a grey squirrel walking in brazen fashion about the garden in spite of Jane who regularly chased it up an apple tree. Then, with increasing temerity it would sit on its haunches near the dining room window grimacing at us as we sat at tea. Dug took a pot shot at it and missed; Jane cornered it in the walled garden and had her cheek bitten through. Meanwhile it continued to enjoy our eggs.

"You'll never believe what that squirrel's been up to now," said Dug one morning at breakfast.

"Try me," I replied.

"He's taken all the labels off the wheat sacks and piled them up in the corner of the meal house."

"He'll have to go," I said.

A few days later he made the mistake of showing himself at the top of the elm tree when Dug had the gun in his hand. He brought him down with a single shot.

There was no more evidence of his mischief about the farm and garden, but we missed him. Even Nigel could not bring himself to claim the shilling for his tail.

Pheasants were a common sight. Over the years the breeds have changed, but then it was the Old English Ringneck that strode stately and colourful about the lawn and drive or sat in rows on the walls of the back garden.

At dusk the raucous cries of the cock birds as they flew up to roost came from every wood. One roosted nightly in the tall cypress tree by the nursery window where the boys could watch him until darkness fell.

Our life was getting organised. Nigel had started school at the old flint-built church primary school in the village and Mary, with Nick, Peter and Jane were becoming familiar figures on their way to meet him each afternoon.

Mary and I had picked most of the apples and carried them, loaded in log and laundry basket, to store in the spare bedrooms and in the cellar until a heady, cidery fragrance exuded from above and below.

In keeping with many houses and even small cottages in the village, the farmhouse had a capacious cellar, used, so it was said, to store contraband goods when smuggling was rife in the area in the 18th century. There were also quaint little protrusions between the bases of the tall chimney in the roof under which there were cubby-holes for the same purpose. As these could only be reached from the inside of the chimney, I never felt the urge to explore them in search of forgotten booty.

We were almost settled in when Jane had her pups. Since her arrival she had endeared herself to all with her trusting affectionate nature and the way she had of drawing back her lips in the friendliest of grins. She had soon become known as the dog with a smile.

Her condition had caused much interest in the village so that I was constantly met with inquiries of, "How is Jane?" or "Has Jane had her puppies yet?" When the event at last took place it caused as much excitement as a royal birth.

The puppies were not pedigree, like herself. We had tried often to mate her, but she would have none of the K.C. registered sires we presented to her, although there had been a misalliance or two with a disreputable Casanova that roamed our Hampshire village.

This time she had done rather better for herself with a handsome yellow Labrador from the Manor House, although I thought they could have chosen a more discreet place for the copulation than the middle of the lane, just as the school bus was rounding the bend. It had stopped and waited while I manoeuvred the pair into a field and shut the gate, then it lumbered away with curious small faces pressed against the windows.

Now, when I could tell that the pups were imminent I made Jane comfortable on the wooden platform surrounded by straw bales we had prepared in the corner of a loose box.

As the boys were in bed with sore throats and temperatures and insisted on my ministrations rather than Mary's I spent most of the morning running to and fro from the stables to the nursery.

Returning to the loose box about mid-morning I heard a faint squealing. Eagerly I looked for a puppy, or puppies, but there were none to be seen.

Jane looked at me anxiously as the squealing came again. This time I traced it to a heap of straw in the corner of the box and found a wriggling pup still in the envelope in which it had been born. Under more straw in the opposite corner was another.

I picked them up and gave them back to Jane. They were getting cold. In spite of my urging she took no notice of them only stared at me in the same pathetic bewilderment and the pups wriggled away again joined now by a third and quickly after, a fourth.

Of course I should have known what to do, but my maternity experiences had been mainly with cows. Whelping bitches, I had found, usually preferred to be left alone and had coped quite efficiently.

It was lucky for Jane's litter that Len, the keeper, came into the yard at that moment to fetch his tractor. I called to him. Putting his head over the door he took in the situation at a glance and grinned, but he came in and picking up the first puppy, did what I should have done, splitting the envelope across the pup's face and taking some of the mucus on his fingers rubbed it into Jane's mouth. She got the idea in one and began to lick the little creature, then the others as Len gave them to her. When four more were born she needed no assistance but we lost the first two. My efforts to warm their cold stiff bodies back to life in the Rayburn were useless. All the same there were six strong, healthy pups.

Later Len picked out a dog and a bitch for himself and a Golden Retriever/Labrador being a good working cross, we had no difficulty in selling the others.

Jane was a good, if somewhat resigned mother. Her comparable age in human years being forty-nine, she was probably a little old for her first taste of motherhood. She never played with her pups as a young bitch does and I think she was relieved to see them go.

One thing that made life interesting at Court Hill was the constant element of the unusual. One day shortly after our arrival I opened the back door to a tall distinguished-looking man, who informed me in cultured tones that he had come to clean the windows.

I was delighted. Window cleaners are hard to find in the country and the windows of the farmhouse were large and many.

This window cleaner was not only efficient at his job but he was a remarkable source of information, on the topography and history of the village.

Did I know, he asked, as he filled the first bucket with water from the hot tap, that Hilaire Belloc had lived in this house?

I am ashamed to recall that at that time my knowledge about that doyen of Sussex verse was extremely vague, a deficiency that was to receive considerable repair in the coming years, starting at that moment.

"Don't you know his poem 'Halnaker Mill'? The mill's near here, you can see it from the piggeries on the Downs."

I confessed I did not.

"Well, it goes…" He took up a stance in the doorway, bucket in one hand, chamois leather poised in the other, and recited in a sonorous voice the lines of Belloc that were to become so familiar to me.

"Sally is gone that was so kindly,
Sally is gone from Ha'naker Mill,
And the briar grows ever since then so blindly
And ever since then the clapper is still
And the sweeps have fallen from Ha'naker Mill".

He dropped the chamois into the bucket and disappeared round the front of the house.

Over the second bucket he talked about the Manor House, which, he said, had been built on the site of an earlier manor that was for eight centuries the property of the See of Canterbury.

He seemed just the man to explain to me the old flint archway that stood against a backcloth of woodland on a hilltop to the west of the farmhouse. When he came back to the tap I had question waiting. Nor was I disappointed.

"You mean the Folly? That was built by a Countess who lived at the Manor, to make work for a master flint builder who was destitute after the Napoleonic Wars, so the story goes. There used to be a tea room at one side, where she took her friends to picnic and admire the view."

He left me at the end of the morning, not only with windows sparkling clean but also with my mind very much improved.

Soon after this, a fall of soot behind the Rayburn indicated the need of a sweep.

'Old Fred', the senior member of the farm staff, said he knew just the man. "I'll be seeing him in the pub tonight, I'll tell him to give you a call."

The next day, a little old man arrived on an aged bicycle to which was strapped all the paraphernalia of his trade.

He surveyed the chimney.

"Aah, tis a good ole wide 'un," he observed with satisfaction. "You don't see many of 'em now. Well, you kin leave it to me. I won't make no mess."

So I left him. When I returned he had taken down the screen and was dismantling his rods.

"There isn't much I don't know about this sort of chimney," he told me over a cup of tea. "Bin sweeping 'em ever since I were eight years old. I climbed all the chimneys in Hampton Court Palace on me first day at work."

"You were one of the little boys that went up chimneys?" I asked in surprise. Next to a window cleaner who recited poetry, what could be more fascinating than a sweep who had actually climbed chimneys.

"I remembers that first morning..." he sipped his tea thoughtfully, "Father said he'd call me at three o'clock. We 'ad t'start at four an' be out of the Palace by nine."

11

He went on to tell me how he went back to sleep after the first call and of the bucket of icy cold water from pump that his father woke him with the second time.

"Bitter cold mornin' it were too. That woke me all right. Then father took me downstairs and dried me and gave me a hot cup of tea and off we went."

He passed his cup for me to refill. While I did so he told me how, after sweeping one chimney he had seen his black face in a mirror and run home to wash off the soot... and of the boxed ears and scolding he received when he got back to work.

"How old d'you think I am, ma'am?" he asked, as he rose to go.

I looked at the wiry frame, the ruddy cheeks, scarcely lined and the merry blue eyes.

"Sixty-five?" I asked.

"Another ten years," he replied proudly. "I'll be seventy-five come next May."

"You must have had a very hard life," I remarked. "How have you kept so fit?"

"Well, we 'ad wot food we could get as youngsters, sometimes it weren't much, but first thing every mornin' we 'ad 'alf a pint of beer with a dash of rum. I was 'avin that afore I was twelve. Well good-day to ee ma'am."

He swung the rods onto his shoulder, picked up his brushes and walked jauntily down the path to his bicycle.

The next time I needed a chimney swept, I looked forward to seeing the old man again, but by then he had retired and two young men came from town and did the job with a vacuum cleaner. It was quick, clean and efficient, but devoid of romance. I told them about the old sweep and his story.

"Oh, did he tell you that old yarn?" they laughed. "That's what he tells everyone. There ain't a bit of truth in it. His old man was a scrap merchant, never swept a chimney in his life!"

Be that as it may, I had still enjoyed the old sweep's tale.

Gradually, I made the acquaintance of the farm staff. John was the foreman. A veteran of World War 2, he was gaunt of frame and rugged of countenance. Having escaped from Dunkirk in a rowing boat and survived many hazardous experiences in the desert, he was unperturbed by anything untoward that might occur on the farm. He was a calming influence when tensions arose and a twinkle was seldom far from his keen, blue eyes. He was an authority on country lore and when the boys were older and developed an interest in wildlife, it was always "ask John" when in doubt. A bachelor, he lived in lodgings in nearby Fontwell.

A close friend of his, Maurice, the head tractor driver, lived in the village with his wife as did 'old Fred', a kindly man with long experience of farm work. Charlie, his brother lived in one of the farm cottages beyond the farmhouse with his sister Fanny and her two sons.

George, the rabbit catcher, a small, thin elderly man and his wife Rosie, lived at North Wood, a cottage in an area of field and woodland with farm buildings that housed a second dairy. Rosie was stout, raven-haired and red-cheeked. The contrast in build from her husband made one think of the old nursery rhyme:

"Jack Spratt would eat no fat,
His wife would eat no lean".

Rosie had two passions; collecting good paintings and books on country life, many of which she lent me. It was Rosie who introduced me to the fascinating writings of Denys Watkins-Pitchford, who wrote stories of wildfowling as 'BB' and to Fred Archer's tales of poachers.

Besides a large dairy herd, fifty breeding sows and bacon pigs, the farm carried five hundred Clun Forest ewes, looked after on the Downs by Shep Wells. Shep was eighty years of age and as upright as a guardsman, but for all that stood no more than five feet high. A ruddy complexion contrasted with snowy white hair and a long moustache. In spite of the fierceness of his blue eyes, he was jovial and warm-hearted.

Long widowed, he lived by himself in one of the farm cottages in the village. He had a married daughter in Bognor who came in regularly to 'do' for him.

Cluns lamb late, not until March, but at the end of February Shep moved with his dog into the shepherd's corrugated iron hut on the Downs. The winter after we arrived at Court Hill was bitterly cold and March brought freezing winds and flurries of snow.

One morning Dug came in to breakfast looking worried.

"Old Shep is not well," he told me. "He looks rough, but he won't go home."

"What's wrong with him?" I asked.

"Flu or bronchitis I should think. He can't stay up there."

He went back after breakfast with aspirins and a cough remedy, which Shep would probably ignore, and more persuasion.

"Don't ee talk such fool nonsense," the old man replied, huddling over the coke fire stove. "How can I take to me bed an' leave ee with all those yows to lamb?"

The next morning Shep was worse. Dug picked him up and carried him to the Land Rover and took him home. Having put him to bed and called the doctor he fetched the daughter from Bognor.

When our local GP arrived, Shep glowered at him.

"Wot be ee doin' ere?" he growled. "I don't want ee, never seen a doctor in me life."

"Well, you are seeing one now," replied our GP, pushing a thermometer in his mouth, which effectively stopped his complaining.

It was the beginning of a long illness. Shep was moved to his daughter's home and we visited him there during his convalescence. He was as perky as ever, telling us about the speedboat trip he had taken the previous summer to celebrate his eightieth birthday. He never returned to work, but went to live with another daughter in Surrey. Some years later she brought him to see us. He had not aged or altered one whit. It was many years later that we heard he had died. He must have been well over ninety.

The fattening pigs were looked after by Bill, in the piggeries on the Downs, next to a large barn where he milled the barley for their feed, also the breeding sows in huts in the open fields. Bill was a volatile little man, known as 'the squeaker', not so much on account of his occupation as for his very high-pitched voice, which rose even higher when he was recounting some crisis concerning himself or the sows – and crises dogged Bill all the time, such as the night he hammered on the back door as I was getting supper, declaring that he had drunk a bottle of paraffin. Having calmed him down a little, I asked to see the bottle, to discover that the drink had been cold tea in a bottle that had once contained paraffin, after which he retired to the pub for his more usual tipple.

He lived some miles away and drove to and from work in an ancient Austin Seven, which he also used for carting a few bales of straw or sack of meal.

One day when returning from a walk along the farm road I saw Bill's Austin weaving slowly toward me, the springs well down on the passenger side. In the front passenger seat I made out a wide pink brow, then two pink ears and finally a quivering snout.

Sitting in the front seat, with great dignity and obvious enjoyment, was a large, very pregnant sow.

Bill pulled up beside me.

"Just taking me girlfriend for a ride," he grinned. "She broke out on the Downs and were off to farrow in woods. I'll put her in one of the sheds to have her litter."

How did you get her in the car?" I asked.

"Oh, she gets in by herself. I often takes her for a ride. Don't I gal?"

"Ee'll 'ave t' wash the seat out afore 'e takes the missus out," said old Fred. "Ever see 'is missus get in the car? She settles herself down like a hen on eggs."

Some time later Bill left and Sam took his place. He was a tall, strong man, solid and dependable, with an insatiable appetite for work and an innate sympathy with animals. Like John, he was a good man to have around in a crisis.

The farm staff changed over the years, especially the cowmen at the two dairies. The characters I have recorded are those whom I will not forget.

Another memorable character was Miss Fanny Smith, who lived with her two sons, Tom and Bill, and her brother Charlie, who worked on the farm, in the second cottage past the farmhouse. A short, plump, rosy-faced woman, with snow-white hair drawn tightly back into a bun, 'Aunt Fan' as she was known in the village, was a kindly, popular character with a certain dignity, probably born of her young days in royal service as a maid at Windsor Castle. She was frequently to be seen, when her work was done, leaning on the garden gate, clad in her starched, all enveloping, cotton pinafore and I would stop on my walks with the dogs and chat with her.

Although she never went beyond the gate she knew everything that went on in the village and beyond, news being brought to her perhaps, by her sister-in-law, Mrs Fred Smith during her daily visits, or by Tom, after his retirement from the Royal Navy, from his nightly jaunts to the Newburgh Arms.

On leaving school, Bill joined the farm staff, working for a while as assistant gamekeeper, then with the pigs, graduating to tractor driver. It was not long after he had taken up this occupation, that Dug, seeing him in the yard one afternoon, called him into the office. After a short conversation, which included remarks about the weather, as a violent, south-westerly gale lashed the branches of the tall beeches opposite, Bill turned to go.

"Just a minute Bill..." Dug said, not quite knowing why he did so. Bill paused and a moment later, in the time it would have taken him to reach his tractor, one of the beeches came crashing down, smashing the machine. Bill would have been killed outright.

2. Winter

The last leaves were off the beech trees, swept into deep drifts by the equinoctial gales. For weeks we had raked them from the lawn and drive and swept them from the porch and even from indoors for they had blown in through every open door and window.

The bareness of the trees opened up new vistas. We could see the lichen covered, tiled roofs of the buildings and across the fields to the dark mass of Eartham Woods.

November was a month of still, misty days and fierce sunsets. The house faced west so that its tall windows looked straight into the florescent glow as the crimson ball sank behind the trees.

Until it was almost dark in the room I could not bring myself to draw the curtains until the last pink flush had faded and only a primrose glow tinged the amethyst sky, the woods darkened into the distance and the evening star appeared above the black silhouettes of trees.

We watched the birds fly in to roost; starlings chattering in the sycamores; wood pigeons in the wych elm; a screech owl in the yew by the potting shed; sparrows twittered in the tall fuchsia by the porch while a wren stole swiftly, like a little feathered mouse, from stem to stem.

As winter approached we found the house less hospitable. The north wind whistled through the well house, rattling and slamming the several doors; it buffeted the kitchen windows and entered every crack and crevice until an icy draught seemed to come from every direction; it sneaked through the ill-fitting kitchen door until my legs froze as I stood by the sink. Even the Rayburn at one end of the big kitchen made little difference to the temperature.

In the evening we were glad to huddle round the big log fire. Leaving it to brave the nether regions to get supper was a test of endurance. For a moment I would stand in a state of shock, like someone having dived into icy water. Then I would dash around, completing the task as quickly as possible and return gratefully to the fire to thaw out.

But even the fire did not warm us when the old house trembled in the teeth of a south-westerly gale. We would sit with rugs round our legs until a mighty gust from the cellar below sent the carpet billowing round our feet. Then we would retire, clutching hot water bottles to seek warmth beneath the blankets.

Mercifully the children seemed untroubled by the cold, even Nigel on the top floor where cold air constantly blew down from the roof, stuck it out.

Over the years a new roof and new flooring in the kitchen made the house more bearable, if never cosy, in cold weather, but while others in centrally heated houses went down with colds and flu, we shivered and stayed healthy.

We never removed the ashes from the fireplace and these retained some heat until the fire was lit again.

When Dug tired of splitting and sawing logs he brought in some old fencing posts to burn. These stuck out across the hearthrug and we pushed them in as they burned away.

After a long day on the farm Dug often retired early, sometimes before the nine o'clock news. Then he would carry out to the drive what remained of the posts, while I ran behind stamping out the fallen sparks, in the happy assumption that he was saving them to burn the next evening. But after this nightly ritual, while Mary and I huddled over the fading embers, the logs outside continued to burn merrily into the night.

By now I was familiar with the geography of the village, which was simply a loop of roads slung from a straight top road running parallel with the Downs. There were a few shops; a butcher's near the Newburgh Arms in Top Road and further along the road a small general store run by Mrs Apps, a kind comely woman whose husband, the scion of a long line of Apps in the village, kept the beer house adjoining.

When the boys stopped to buy sweets on the way home from school, Mrs Apps always had a dog biscuit for Jane, so that, in time, the dog refused to pass the shop, sitting down solidly until she had her way or was dragged home.

Near the pond was a licensed grocers, an old fashioned shop, the door approached by a flight of steps. Inside were two long counters one either side of a narrow aisle. In time it was modernized and made self-service, without losing the atmosphere of the village shop.

It was there one met one's friends and discussed the events of the day. In those far-off days when petrol was cheap and time more expendable, we were well served with tradespeople who called: bread was brought to the door; a butcher brought meat twice a week; the grocer called for orders on Tuesday and delivered on Friday; the cobbler from a village some miles away took shoes to be repaired one week and brought them back the next. From a shop in another village that sold everything from haberdashery and knitting wools to hosiery and boots and shoes, the proprietor brought out samples of his wares. The calls were mostly made in the evenings and a cup of tea or coffee would be consumed while relating news items from the surrounding villages. A fish and chip van called in the village on Friday nights and sometimes Dug could be persuaded to fetch some for our supper.

We were now regular attendees at St. Mary's Anglican church, which had its origins in the 12th century but had subsequently been enlarged and later restored in Victorian times.

The benefice was united with that of the neighbouring villages Eartham and Madehurst. Our incumbent was rector of Slindon but Vicar of the other two Parishes. Of the church-wardens, one was a magistrate whose father and grandfather had been Rectors of the church and the other a retired naval Captain, indeed so many of the residents at that time were retired naval officers that the village was nick-named 'HMS Slindon'.

There was also Mr Gregory, the verger, who carried out his duties with care and dedication. Looking older than his years from a lifetime of working as a gardener, he was bent of back and knee and splayfooted.

"He walks down the aisle as if he was treading his celery trench!" someone was heard to remark. When he retired and moved to Hampshire he was sadly missed.

November brought the shoots, which were held on the farm on alternate Saturdays. Dug went with them as back gun. Beaters assembled in the yard after breakfast, then the guns. By 10 o'clock the place was deserted and silent, except for the occasional sound of distant gunfire. I felt singularly lonely after Mary had taken the boys for a walk, carefully in the opposite direction from the guns, until Mrs Long arrived with her housekeeper and took over the kitchen to prepare a hot lunch for the shooting party, which they consumed in the dining room. Often a dog or two would sneak in with them and on occasion, I found a wet spaniel in my bedroom. At the end of the afternoon when light was fading, guns and beaters returned, rows of pheasants were hung in the well house until Dug took them to the butchers.

Some years later a shooting lodge was made and furnished in the unused stables, with its own kitchen and cloakroom, a more satisfactory arrangement for all concerned.

Despite the discomforts of our domain we managed to persuade two of Dug's most intrepid relations, his youngest sister and her husband, to spend Christmas with us, warning them to bring plenty of warm clothing and their own hot water bottles. Not only did they survive the experience but came every succeeding Christmas for many years and the festive season took on, for us, a pattern it was it was to follow for a long time.

Having completed, many weeks before, the preliminaries of making puddings and Christmas cake and the present buying and secret wrapping, the Sunday before Christmas we took the Land Rover and filled the back with holly and other evergreens. The woods and lanes were well stocked with those three native trees of Sussex, the beech, the yew and the holly and there was always a bush or two of the latter to be found, laden with berries.

During the following week Mary and the boys decorated the house and then the tree, which always reached to the ceiling. I helped Dug saw and cart a good stock of wood to keep our fire going over the holiday.

Our guests arrived early on Christmas Eve in order to help with the children's party. Then the old house reverberated with the sound of running feet, squeals and cries as the old-fashioned games of sardines and murder and hunt-the-thimble were played, full use being made cupboards and various nooks and crannies.

When the last child had gone home and the party debris cleared away, we left our long-suffering relatives putting the boys to bed, Mary having gone home for the holiday, while we went to a Christmas Eve party that was held in the same house every year in the village.

Returning home, hoping that the boys would be asleep, there were the stockings to fill, gifts to put round the tree and all those last minute preparations for the following day.

Then came the hours to fill in, talking or listening to the radio, until it was time for midnight Mass.

On Christmas Day the boys were up at 5am, most of their presents were opened and many sweets consumed before breakfast. By mid-morning, preparation for the festive meal having been made, we all went to church, then in the afternoon, a walk over the Downs.

The Christmas always ended, on Boxing Day, with a visit to an ice show in Brighton.

The next day our brave guests departed and life settled down to the normal routine.

A belated Christmas Present from Dug to myself was a second golden retriever. One day early in the New Year we drove back to Hampshire, to a village near Selbourne, from whence we had come three months before, to a breeder who had a half-grown bitch for sale. Her name was Beauty of Anbria. I nursed her on my lap, large as she was, back to the farm where she was rapturously greeted her by the boys and suspiciously, but not unkindly by Jane. The two soon became friends and with the younger dog, Jane who had been prematurely aging seemed to get a new lease of life.

The back door slammed and I heard the thud of gum boots landing on the floor of the wash house and the rustle of oilskins as Dug took off the second lot of wet clothing that day. The rain had slackened but the wind still buffeted the windows in angry gusts.

Banishing the dogs from the fire I put on more logs and drew up Dug's chair.

"What a day!" he exclaimed as he came in, dry clad once more. Stretching out his legs to the blaze, he picked up the newspaper while I repaired to kitchen to get supper.

Immediately there was an urgent hammering at the front door.

"Oh no, not tonight!" Dug groaned, tossing aside the paper and hauling himself to his feet.

I went to the door.

"He's here again," called Len, as if I had not guessed.

"OK Len," came Dug's resigned voice, "give me a minute."

"What about supper?" I asked.

"Oh, just fix me a sandwich and one for Len."

Back in the washhouse Dug struggled into clammy gum-boots and another mac, half dried from this mornings wetting, while Len dripped in the hall.

"Sort of night he likes," he observed, "wet and windy, the birds sit tight."

'He' was a poacher who had been busy in the woods all winter. Not a casual poacher, popping off the odd bird for the pot, but a professional who made sufficient income from selling pheasants to hotels to keep him the whole year round.

Dug and the keeper had spent many nights in the woods, in all weathers, sometimes accompanied by Stan, the local police constable, a stout, amiable man but a stern upholder of the law.

"Your poor master," I said to the dogs, "this is the fourth night he's been out this week."

They drooped their ears in sympathy while sidling back to the fire. Often Dug was not back until nearly dawn but this night it seemed that I had not been in bed long when I heard the Land Rover returning and roar into the yard.

"Did you catch him?" I asked, as Dug came into the bed-room.

"No. Wait a minute and I'll tell you." He climbed into bed and warmed his cold feet against mine.

"We were up in Little Wood, we'd been there a couple of hours and seen nor heard nothing, then Len thought he saw a light in the house and we decided to take a look. I'd climbed over the barbed wire fence and Stan was half way over when Len let go of the wire further on. It caught Stan smartly between the legs and lifted him into the air – you should have heard him! Or maybe you shouldn't, his shouts and curses could be heard a mile away! It was no good hanging around any more after that."

I giggled. The thought of our portly constable perched on a strand of barbed wire was irresistible.

There was little light heartedness in the following weeks as other men from the farm joined in the hunt. They became weary and tempers frayed, but still they were no nearer catching their quarry, although they knew who he was and where he lived. More than once they tracked him to his door, but when police entered, he was asleep in bed and no trace of his booty could be found. Eventually, a well-planned offensive was launched, with the keepers and men of the two bordering estates taking part, and many of the West Sussex Constabulary. When I went to bed I could see lights flashing on the Downs and answering signals from the fields near the house. They were closing in.

At about 1.30 am I was awakened by the sound of cars and Land Rovers pulling up in the yard, of doors slamming and voices. Many footsteps converged on Dug's office and the door was shut.

It seemed they had caught their man and had him in the office. The voices went on for a long time, sometimes raised.

I wondered if I should go down and offer them all coffee, but considered that a woman in a dressing gown might not be welcome. Soon afterward, the gathering broke up.

There was more shuffling of heavy boots and banging of doors, then voices outside, noisy revving of engines and slamming of car doors and, at last, silence and peace.

"Thank goodness that's over," said Dug when he came to bed. When the poacher appeared before the local bench he was allowed bail. It was all very amicable. Over a drink with Dug and Len afterwards he told them how he had been crouched in the undergrowth for two hours one night while the two of them had waited not more than a yard from him. One could not resist a grudging admiration for the man's stealth and temerity.

At the quarter sessions he was sentenced to three months imprisonment, which seemed little enough for all the birds he had taken and the trouble he had caused, but he bothered us no more and the following winter conducted his illicit activities on an estate in the next county.

3. Spring

One morning early in February I looked out of the window and saw them, their glistening white spears pushing up between the couch grass clumps on the wild bank in the front garden – the first snowdrops. I went out to look at them. They were short stemmed as yet, clustering tightly together as if afraid of the cold. In a few days they would be tall and standing free, hanging their dainty green-veined heads.

Then the aconites, which vied with the snowdrops as the first harbingers of spring, raised their yellow cups among the grasses. By mid March the whole bank was golden with daffodils, and in the wood beyond the cottages were primroses, white and blue dog violets, the dainty muschatel and a constellation of star like wood anemones. The fluffy white blossom of the blackthorn was out, bringing with it the bitter winds again – the blackthorn winter – and the long golden tails of the hazel catkins.

Spring brought great domestic activity among the birds in the walled garden. Blackbirds and thrushes built in the holes and crevices of the flint walls, a blue tit in the stack head of the water pipe by the window and a wren in the yew tree – perhaps the one that had come in from the December cold and roosted for three nights in our Christmas tree.

Chiff-chaffs appeared darting among the spring foliage and the swallows gathered mud from the farm pond for their nests in the stables. Presently their offspring came to sit along my clothesline like a row of diminutive white-ruffed choirboys. They came in through open bedroom windows and daily I went to the aid of one unable to find its way out. One morning while waiting for the copper to fill, I was fascinated by the courtship display of a male chaffinch on the bird table; the fervid head ducking, the vibrating wings, the fanned tail. Meanwhile the object of his affections, a disdainful creature, peaked unconcernedly at crumbs, until, bored with the whole affair, she flew away. This was lucky for me, for only then did I notice a pool of water round my feet from the steadily overflowing tub.

Some birds chose strange nesting sites. A blackbird built on top of Dug's golf clubs in the summer house, the base woven firmly onto a driver, and hatched a brood. Another made her nest on top of a post by the door of the workshop where the men constantly passed to and fro, while a wren constructed his intricate little home in a coil of rope, the top of the coil forming the roof.

Pheasants walked their broods up the drive; the charming little red-legged partridges also brought their young and seven mallard ducklings swam on the farm pond. Lapwings laid their eggs on the ground and the tractor drivers carefully moved them from the path of their machines.

All winter a moorhen had frequented the walled garden, feeding with the hens or with the wild birds, her large feet, like great black spiders, looking incongruous on the bird table.

"How will she find a mate in the garden?" asked Nigel, always concerned for the welfare of our bird life.

"Nature will take care of that," I replied, and sure enough one morning there was a second moorhen scurrying furtively among the bushes, his broad tail showing its white under feathers. For a week they ran to and fro in the garden with their swift agitated gait, then he, perhaps not wishing to continue his courtship under the inquisitive gaze of the hens, took her off to the farm pond. In the weeks that followed I often saw them as I walked or cycled down the hill and rounded the bend, scurrying across the road, long twigs or wisps of straw trailing from their beaks. The nest on its carefully constructed platform grew steadily, half hidden by an overhang of dried vegetation. A few more weeks and there were five tiny blobs of dark fluff on the pond. Moorhen chicks always reminded me of Silas, a black Labrador of great character Dug and I had when we were first married. The cottage where we lived overlooked a pond where every spring moorhens nested. When the chicks appeared Silas would sit and watch them. Then he would take one from its home pond and carrying it carefully, the little beak sticking out from one side of his mouth, the feet from the other and deposit it on another pond two fields away, returning for another chick until he had transferred the whole brood. He would then sit and watch them swim for a while and take them back again. Soon even the parent birds grew accustomed to this procedure and were unconcerned.

One evening I bumped into a screech owl as she flew out from the yew tree. I don't know whether she was more surprised than I as her wings brushed my forehead. She spent so much time in the yew or on top of the summerhouse contemplating it that I hoped she intended to nest there, but perhaps the sight of Jane or Beauty persistently ratting by the tree put her off and after a few weeks she left us.

There were many birds that we saw in the woods or on the Downs that did not come to the garden. Charms of gold-finches feeding on the thistles, bullfinches and wheatears, siskins in the beeches in the lane and crossbills that lived on pine cones in the conifer plantations.

Living in this environment, it was not surprising that the boys became enthusiastic naturalists. First came the pressed flower collections. Nigel's for school numbered over three hundred species. The egg collection, though only garnered from abandoned nests, soon filled a small cabinet. After that came bird spotting. If we could not identify a species from our growing library of reference books, there was one man on the farm who never failed to do so; Dug's headman, John.

John had been born and raised in the village and knew every inch of the woods and Downs and every species of its flora and fauna. As a lad he would climb the tallest tree to look at a bird's nest.

His keen blue eyes saw the first swallow on the Downs, a hawk moth scarcely discernable against the bark of a tree, a Jack heron flying over the fields from Chichester Harbour to, perhaps, a place on the Arun. He knew where a pair of buzzards nested in a holly tree on Nore Hill and where a bee orchid grew and where a roe deer had left her fawn.

All these things he pointed out to the boys.

This was a good place to bring up a family.

By April the lambs had left their mother's flanks and played in groups, like children at a kindergarten; gambolling by the fence, pirouetting on an ant hill or chasing each other round a group of birch trees.

Shep Wells place had been taken by another shepherd, assisted by his sister, Ethel. The new 'Shep' was of a round and smiling countenance, the colour and texture of a well ripened, rosy apple. His eyes, set in a network of wrinkles on his deeply tanned face, seemed intensely blue.

Contrasting with her brother's merry nature, Ethel was timid and taciturn, her long, melancholy features seeming pale beneath their tan. She was always working and walking, carrying sacks of cakes for the sheep, bales of hay, fleeces, clad always in a long, shapeless raincoat, enveloping her thin frame, an old beret pulled down over her straight grey hair and heavy gumboots on those tireless legs. Even on the long walk home to the cottage she shared with Shep, she bore a bundle of kindling. Yet she would never accept a lift in the car or Land Rover, from the village shop or home at night, seeming only to trust her own two feet. When Shep was provided with a pony and cart for transporting the fleeces Ethel always walked behind.

She seldom responded to my greeting, scurrying away like a frightened field mouse, except at lambing time. The birth of the lambs changed her personality; her face radiated the joy of a mother with a newborn child. She met me then with a broad smile and chatted happily about her charges. I made the most of the pleasure of talking to Ethel on these occasions, for it would not happen again until another spring and another lambing.

Shep, however, became unusually reticent at these times.

"How many lambs have you got Shep?" Dug would ask.

"A few," would be the careful reply.

"Many twins?"

"Aye, a few."

"Many left to lamb?"

"A few."

And with that Dug had to be content.

Easter, like Christmas had its special routine, starting with gathering primroses on Good Friday for decorating the church. Dug would take us in the Land Rover. Turning by the farm pond, we rattled and bounced up the rough, flint-strewn lane. Primroses grew here under the hedgerows but not enough for our purpose. When I wanted a few for the house I would scramble on my knees beneath the thorn bushes to reach the longest stems, but now we went on past the big barn to the isolated cottage where Shep and Ethel lived and turned right, along the bridle path flanked by high hedges, where we had gathered holly at Christmas.

Just before we reached the forestry enclosure we drove into an ash wood. Here clumps of primroses grew as big as foot-stools. That first year we filled a laundry basket without showing where we had been. A whole evening was spent bunching them ready to take to church on Saturday morning to be laid on the windowsills and round the font and lectern.

On Saturday evening, a service was held at the tiny church in the neighbouring hamlet of Eartham, the service of the Pascal candle. This was once a pagan rite, originating in Greece, where at this time of the year, all the fires in the cities were put out save the one at the high altar, to be relit by runners bearing torches lit at the altar fire.

The ancient church of St. Margaret's Eartham was a perfect setting for the Christian interpretation of this ceremony. Unlit candles were placed in its many small windows and in the alcoves on either side of the chancel arch. There was just sufficient light to see the massed daffodils and sprays of forsythia. Presently the verger, ignited a taper from the altar candle and followed by the Rector, lit the others placed about the church. It was a short but impressive service.

The climax of the holiday was the annual point-to-point races at Cowdray Park on Easter Monday. It was my favourite day out of the year, starting with a drive through some of the prettiest West Sussex countryside to the Cowdray Estate, an expanse of common, park and arable land with beautiful red soil and a picturesque lake.

Because the Cowdray Hounds hunted over the farm we received a complimentary ticket for the Farmer's enclosure, where, having arrived early enough to secure a place near the rails, we spread a rug, picnicked and relaxed and enjoyed a good view of the racing.

But because young boys can never keep still we worked off some their energy with a walk round the course and an exploration of the ruins of the old Cowdray House, which are open to the public.

There is an aura of romance and tragedy about the gaunt shell of what was once a splendid edifice, so that one can well believe the story of the curse that was fulfilled in the destruction of the house by fire in 1793.

'The Battle Abbey curse', as it is known, allegedly dated from the dissolution of the monasteries. Henry VIII gave to Sir Anthony Browne, whose descendants became owners of

Cowdray with the title of Viscount Montague, the abbey and all its lands, whereon the last Abbot of Battle, on leaving, warned the despoiler that by fire and water his line should perish. A few days afterwards a fire destroyed his home and his heir was drowned while attempting to shoot the rapids at Laufenberg. Later the two sons of his sister, who inherited, were both drowned at Bognor. Many historic relics were lost in the fire, including the sword of William the Conqueror, his coronation robes and the list of barons who accompanied him to England.

Strolling round the remains of the great halls and court-yards it is easy to picture the house in its heyday, when Elizabeth I was entertained there with music and feasting and deer hunting in the park and slept in the velvet bedchamber.

Far removed from thoughts of curses and sudden death are the crowds on Easter Monday, enjoying the sport in the spring sunshine, its light enhancing the bright hues of the rider's tunics and caps and the sheen on their mounts as they are led round the paddock or prance to the starting point.

There are, of course, less clement days, when pelting rain makes a quagmire of the ground and the winning horse is the one which can stay on its feet. At the end of the day, drivers view each other's spinning wheels and wonder if they will ever get out of the car park mud.

The fifth month of the year is perhaps the most splendid in the countryside. While spring brings a dream of promise – the gentle prettiness of a young girl who has not yet attained the full grace of womanhood – May bursts into a full, lush, almost opulent beauty.

While the beech trees still wear their young pellucid green, the sticky buds of the chestnuts and the silver candles of the white beam open further each day. In the garden lilacs and laburnum, apple blossom and flowering cherry bloom side by side.

Dug disliked the full blossoms of the ornamental cherry.

"Like fat women in pink blouses," he would declare, preferring the daintier blooms of the crab apple and wild cherry of the woods. But at this time of year, when Stanes Wood is a haze of bluebells, unbelievably fragrant in the early morning, the lanes edged with the white lace of cow parsley and the verdure of field and wood makes even a sunless day bright, I can criticize nothing; it is enough just to be alive.

The nightingales were in full song in North Wood, which leads out of the Stanes, so one evening I took Nigel along, hoping to hear them. It was dusk and the wood was full of shadows and the rustling of nocturnal creatures going about their business. A heavy cock pheasant, veteran of many a shooting season, lifted off from the young ferns like a jumbo jet and flopped onto his roost; a woodpigeon, suddenly disturbed, flew out of his tree and along the ride, a whistle in his wings; the white shadow of a barn owl drifted over the field beyond.

We heard the bark of a roe deer and the tittering of various small birds settling in their dormitory but not a note of the song we were listening for.

Len came out of the wood on his way home after shutting up the young pheasants.

"Come to hear the nightingales?" he asked. "They're quiet tonight. Never do oblige when you come specially to hear 'em."

I was afraid that he was right and that we had better get home before it was dark. Then from a low bush we heard a few notes of a singular sweetness and after a few moments silence a rich trill of song from the other side of the wood… but no more, although we waited.

"Never mind," I said, when we were obliged to go home. "We'll try another night."

At 3 o'clock in the morning, brilliant moonlight and a beautiful song awakened me. Philomel was pouring out his soul not far from my window, but Nigel was sound asleep.

Even less co-operative was the albino blackbird whose territory was a clearing in Stanes Wood. Many times I saw it flying from trees to field hedge, looking like a small white dove, but when I took the boys to see it, it never obliged by appearing.

Through the Stanes and sometimes on to North Wood was usually the route of my daily walk with Jane and an excellent place for gathering kindling wood. I learned wooding from a huntman's wife in the New Forest, thinking at first that there was nothing to learn until I collected for her what I considered to be a basket of good starting sticks. She had glanced at them with disdain.

"They're no good, they're all green," she said.

A stick must snap clean if it is good for burning; a green one bends or leaves an oblique break. I could usually tell by the lichen that had gathered on the dead stick, but it must not be rotten. After years of wooding I could tell the best sticks, usually beech, at a glance.

By May I had, by virtue of joining the local Woman's Institute, met a number of other mothers with children the same age as my own and there ensued a series of tea parties, birthday and un-birthday, at our various houses. I found no problem in entertaining our young guests, for that year we had twenty motherless lambs penned on the front lawn. Shep had lost a few ewes that year. Although Ethel came twice a day to feed them, when there were children to tea she allowed them to give some of the lambs their bottle. It was a greater success than anything I've ever organised for the young. It was all too much for Nicholas.

"Are we going anywhere, or is anyone coming to us?" he asked one day.

"Not to today," I told him.

"Thank heaven for that! Now I can do something sensible," was his reply.

Well, you can't please them all.

4. Summer

I remember the summer of 1953, our first at Court Hill, for three things; the Queen's Coronation, the appalling rain that quenched the fireworks and bonfire held on the piece of green opposite the blacksmith's and the whooping cough which kept us in quarantine the whole summer.

The boys got the bug one by one. As soon as Nigel was clear, Nicholas went down and then Peter. What with that and the weather, we did not get down to the beach as we had hoped, neither did I fulfil the plans I'd made for the garden. Eventually I turned my attention to growing things indoors. Somewhere I read an advertisement which said that there was money to be made from mushrooms, even on a small scale. 'Why not grow them in your cellar' it suggested.

Why not indeed. There it was, large and empty and fitted with big two-tier trays just right for the purpose.

I fetched buckets of soil from the compost and spread it out on these, having first sent away for the mushroom spores, which arrived in due course together with instructions for cultivation and a list of markets for my produce.

I sowed the spores, paying careful attention to temperature and moisture and waited. Mushrooms, I thought, came up overnight. In a dark cellar they should be instant, but after a week nothing had happened.

I sowed a second lot, watered them and waited again. Nothing. The following week I received another order form from the firm for a further supply of spores, another list of markets and a leaflet of delicious looking recipes for using the surplus – but still not a mushroom had shown its head.

By this time I had given up all idea of big business, just a few tasty buttons for breakfast would be great. A little mould, but not a speck of white showed itself upon the smooth soil. Then one moist morning in July I looked out onto the lawn. Near the bank was a fairy ring and in this darker longer grass were clustered groups of mushrooms, or were they toadstools? I went down for a closer look.

There were a dozen or more beautiful mushrooms with delicate pink grilles and milky suede caps. I gathered them in a basket and we ate them for tea, cooked in milk and butter.

The next day I removed the soil from the cellar and left it once more to the mice and spiders.

After that atrocious season, 1954 was going to make amends with one of those long hot summers one only dreams about, but we were not to know that, and after several cold wet holidays in Wales Dug and I decided to go abroad in search of the sun. So on a late June day, when the thermometer outside the piggeries on the Downs registered 93 degrees Fahrenheit, we set off in a hot and stuffy train on the first lap of our journey to Majorca, leaving the boys in the care of Mary and my mother. On the plane we got into conversation with a young man who turned out to be a farmer and while I sat gazing in rapture at the fantastic cloud formations in the scintillating moonlight, he and Dug discussed hay and milk yields. Oddly enough, he was bound, like us, for Deya and as there was only one pension in the tiny village where else could be staying but at the same one as ourselves.

So it turned out that although we had breakfast each morning sitting on the patio under trailing blue plumbago and crimson bougainvillea, drinking fragrant coffee out of cups as big as soup tureens and eating delicious yellow bread with goats milk butter, the trend of conversation was the same as at home.

The owner of the pension was an Englishman and received a London paper daily. Out of a mistaken kindness he passed it on to Dug and his farmer acquaintance and they read with increasing gloom of the ill effects of the prolonged drought at home and the damage done by the subsequent heavy thunderstorms and commiserated with each over the state of the crops.

But we did enjoy the swimming and the sun, the trips round the bay to the fascinating little port of Soller and the magnificent display of national dancing that we saw there. The pension was a twenty-minute walk from the sea, so the brochure had said; it even mentioned mules for hire. I rather fancied riding a mule down to the beach, but the only ones we saw were working hard for local farmers. So we walked daily down a precipitous track, jumping from stone to stone across streams and leaping small ravines until we became as agile as the sleek brown goats that watched us with a supercilious air from the ledges of the rocks.

There were herds of black pigs too, that rushed at us along the narrow path. But the water, when we reached it, was worth the trek, even if the toil back to our lunch and siesta was a little arduous.

We discovered on the first day, when we walked up to a deserted village that it really was true that only 'mad dogs and Englishman go out in the midday sun'. All the locals had their heads down behind green closed shutters, but we did find one little café open where we were served tea with goat's milk in a shady courtyard. In the evening the little village came to life. From then until midnight, boys on Lambrettas raced up and down the steep, narrow lanes, women scrubbed their laundry at the communal washing place and old women sat in doorways making lace. On one occasion we saw Robert Graves, the author and poet, a local resident, driving his Land Rover with his several children around him.

Despite the joy of sunshine so constant that if a cloud no bigger than a man's hand appeared on the horizon the tourists frowned, fearing they would not get their money's worth, I soon tired of the arid climate and the sight of green trees and fields as we flew over the English coast towards London was a renewed delight.

However good a holiday has been, the best part for me is always coming home. I return to my nest with all the instinct of a homing pigeon. It was so good to see and smell the Downs, to look round the garden, to see the boys and the animals.

There's a funny thing about dogs – leave them a few hours, a day, a week and they cannot contain their rapture at your return, but after an absence of two weeks or more they will condescend to no more than a flick of the tail and a disdainful glance which says, "so you're back, you needn't have bothered, I was doing very nicely thank you."

The day after our return we took the boys up to Bignor Hill in the Land Rover. It was one of our favourite excursions, entailing a four-mile ride over rough downland tracks between beech woods on one side and open fields on the other. Rabbits, hares, squirrels and sometimes a roe deer ran across the track; goldfinches fed on thistles, and butterflies – tortoiseshells, peacocks and meadow browns – settled on the wayside herbage.

Before the cultivation of the Downs during World War II they had been an open expanse of turf and furze where sheep had grazed and drovers had brought their cattle from Portsmouth and Chichester on their way to London. An old cattle pound still stood on the outskirts of the village.

Deer were still plentiful, roe deer in the lower woods and fallow on the higher ground. During our first winters a white doe frequented the paddock behind the buildings. At dusk one spring evening we took the boys to see fourteen does, which had stolen from North Wood to nibble at the young lays. These were shadowy ethereal shapes against the dark wood, but to see a young buck break cover in full daylight and bound away in front of us, the sun glinting on his reddish hide and bobbing white stern, as it did one afternoon was a pure delight.

Bignor Hill is a wide expanse of heather and furze, fringed with woods and little coppices of dogwood and spindle. On a grassy knoll stands a Roman signpost, renewed from time to time, it still retains the old names *Regnum* for Chichester and *Londinium* for London. This point commands a grand view across the valley while from a higher aspect to the south can be seen the entire coastline form the Solent to Brighton, and the distant shape of the Isle of Wight. Climbing to the North along a chalk track between fields of barley we reached yet another viewpoint from, which we could see the Arun Valley and, in the distance, Chanctonbury Ring. An exultation of larks rose from the cornfields as we passed and from the open down we heard the bubbling notes of a curlew. We could have walked for hours along the sheep tracks and bridle paths, which were fringed with the vivid blue of viper's bugloss, pink centaury, tall yellow mullien and dainty harebell, while in contrast the ubiquitous ragwort reared its brassy head.

There were many ways in which to spend our leisure in this corner of Sussex, had not leisure time been something we usually lacked, but sometimes on a summer Sunday afternoon we would tear ourselves away from the farm to spend a few hours watching the polo at Cowdray Park. The game was fascinating and often exciting. It was pleasant to take a picnic tea in such beautiful surroundings and to stroll on the ground, treading in the divots between chukkas.

Our favourite spot along the coast was Climping, or to use an older spelling, Clymping, between Littlehampton and Bognor. Its long beach of firm sand with a little shingle, on the brink of woods and farmland, had its delights at all times of the year. In summer the gently graduated depth of the sea was safe for children's swimming, while the high breakwaters afforded shade from the sun and shelter from the wind.

Patches of bright stonecrop carpeted the shingle and yellow sea poppies hung their delicate heads; sturdy sea heather and sea kale sprouted from between the rocks. There was always something to watch on the water; yachts and speedboats from Littlehampton, water skiers and fishermen.

In early autumn when mists hung over a calm opalescent sea and the tall tamarisk bushes sported their pink plumes, the water was often still warm enough for swimming or we could look for little green crabs in the rock pools.

Even in winter we enjoyed a brisk walk along the sands under a canopy of screaming gulls and back along the field paths where the thorn bushes leaned, flat topped, away from the prevailing wind.

One January day we were returning to the car when a helicopter from Thorney Island flew low overhead. The boys waved to the pilot, who immediately turned his machine and hovered for a minute or two, less than twenty feet above our heads, the draught from the propellers blowing Peter completely off his feet. The boys stood waving until he was far over the water; the episode had made their day.

This little seaside gem has increased in popularity over the years, discovered by visitors from London. At weekends the beach and car parks are filled to capacity and the wild flowers along the shore have long been obliterated by the tread of many feet, otherwise it is little altered.

Give me a day of perfect conditions at Climping and Spain can keep all her beaches.

On weekends when the pressure of work kept Dug at home, I could always find plenty to do in the garden. Birds are ungrateful creatures. From early summer they rewarded my winter care of them by persistently raiding the fruit and vegetables. One blackbird sat all day on a post waiting for the loganberries to ripen; thrushes took all the cherries and happy bands of finches stripped the redcurrants; sparrows ate the peas and I even saw two hen pheasants working systematically up and down the rows, pecking at the best pods.

I complained to Len about it.

"Aah, they will," he replied with a grin and left me with my problem.

Returning one day from the beach to find every one of the hundred sprout and broccoli plants I'd put in the day before stripped to pathetic little skeletons, I decided on a full-scale offensive with every deterrent I knew and waited to see what my feathered friends-no-longer would make of them.

One morning, looking out of the kitchen window, I saw a young cuckoo perched on the wheelbarrow, which I'd left on the path. He was puffed out like a ball of striped fluff. His wings were still rounded and as yet he had no tail. He must be about eight weeks old, I thought, and flying would still be difficult, landing even more so. I wondered how he had managed it on the narrow edge of the barrow. I wondered what nest he had come from and where and whether his foster parents would come to feed him. I managed most of my tasks with one eye on the window. I wanted to see him take off. Then I was called to the telephone and when I hastily returned he had gone. He is the only cuckoo, young or adult I have ever seen.

The greens were doing well, the birds seeming to at last respect my defences and leave them alone. Then suddenly something was at them again. It wasn't birds or caterpillars. I asked old Will to come and look at them. He had dug the garden in the spring and called in now and again to give advice. Most of the time he caught rabbits and did odd jobs on the farm.

Although known as 'old' Will he was a man of indeterminate years. His face was creviced and lined but his eyes were young and his longish brown hair untouched by grey. He always wore the traditional cords and an old grey jacket that would have fitted a much larger man.

He looked at the cabbages and stroked his stubbled chin with a gnarled finger.

"Aah, tis the bub. You wants t'watch out fer 'ee. You'll 'ave no cabbages else."

"The bub?" I enquired.

"Aah, tis a floi. You wants t'put ashes on un. Not hot mind." He obviously thought I was a woman of very little brain. "When t'dew's gone off, spread it all along t'line. Then you won't get no more bub."

There is nothing like living in a beauty spot in the country, especially in a good summer, to encourage your friends and relatives from far and wide to visit you. That second summer at Court Hill I was amazed to discover just how many we had, in fact I began to wonder if we had been listed in the AA book by mistake.

To avoid turning out of our bedroom for our frequent visitors, we decorated and fully furnished the rooms at the top of the house so that we could, and sometimes did, sleep six children up there.

Not everyone came to stay. We had innumerable callers, from long lost friends to casual acquaintances and fellow employees from previous farms whom we had almost forgotten. Somehow they all looked us out.

Dug was always delighted to see them. He had such an insatiable desire for company that it surprises me we didn't take someone with us on our honeymoon.

I like company, making new acquaintances, meeting old friends, but I prefer a little warning before they descend upon me. Surprise visits usually occur in the middle of a meal, when I am getting down to some gardening or typing I have been trying to do for weeks or just about to sink exhausted into a chair for a few quiet minutes with the daily crossword.

Somehow I could never match the rapturous enthusiasm of Dug's "Look darling who's here!"

I was often reminded of an essay of Madame Belloc's in which she describes a visit with a mutual friend to Marion Evans, the writer George Eliot.

"Marion had a headache," she wrote, "and received us with an air of resigned fatigue." I always hoped that my 'resigned fatigue' was not as obvious.

In the middle of this influx Mary decided to leave us for work nearer her home and in July departed. Perhaps I would get a school leaver to replace her at the end of the term, until then I was 'on my own'.

It was amazing that when a mere slip of a teenager had said "bed boys" my younger two had said "goodnight", ascended the stairs in an orderly manner and once in bed little more was heard of them. Their reaction to my own "bed boys" was a shriek of laughter, a mad race round the house and pandemonium.

Their big bedroom on the first floor had obviously been used as a nursery before. A stout iron bar was fixed across the bottom of each window, but having seen Nicholas's blond head hanging over the sill, I decided that this was not protection enough, so Dug and I fitted three curtain rods to each window. That evening I left the boys with greater confidence and settled down to some writing.

All was quiet for a while, then suddenly came shouts of glee from upstairs. From my desk I saw a bright flash of metal descending past the window, followed by three more.

They stuck, quivering, in the ground like Red Indian spears, to be followed immediately by a white mass of bed-clothes, which draped themselves upon the rods like a row of little wigwams among the rose bushes.

So much for my safety rods.

Nigel, now in his third year at Primary school, had, to some extent, put away childish things and looked with disdain on his brother's bedtime pranks, but was not above joining in with a free-for-all fight at mealtimes or a bit of quiet mischief of his own.

Daytimes were scarcely better than the bedtimes. Immersed in chores of house and garden, I decided to leave Nick and Peter to amuse themselves, which they did with all the inventive imagination of three and four-and-a-half year olds. Wherever there was oil or grease, they found it, and finally black car paint, until I wondered if it would be easier to colour over the bits of themselves that had missed and leave them black. Nightly I put them to soak in one bath and their clothes in another, while the ironing was something I hoped to catch up with by Christmas.

"Leave the work," said Dug, one day when Nigel was at school but I had just rescued the other two from the sludge at the bottom of an old silage pit and cleaned them off. "I'll take you all up on the Downs and you can walk back – tire them out".

Dug dropped us by the sheep, saying we could walk back for a mile or so then he would pick us up by the two Nissen huts on the homeward track about an hour later.

The huts were used for storing the combine harvesters, but that day they were empty. My short investigation having satisfied me that the boys would come to no harm there, I sat down on the grass and relaxed in the sun, leaving them to their game of running in at one end of the sheds and out at the other.

It was sometime before Dug arrived, but the time passed pleasantly. Bees droned in the honeysuckle and somewhere a chaffinch sang; a tree creeper stole like a little feathered mouse up the bark of an oak tree. My thoughts turned to an article I was writing for a nature magazine. The boys' shouts and laughter came through to me remotely. I was glad that they were enjoying themselves so harmlessly.

At last the Land Rover came to a rattling halt beside me.

"Good Lord, look at those children!" gasped Dug, as he got out. I looked into the shed from whence the cries were now coming. There in a dim corner was a pile of sacks of coal, which I had overlooked but they had not. They were climbing the sacks in great glee, throwing over each other handfuls of loose slag, which had trickled out in little heaps.

They were as black as sweeps and I winced at the thought of the cleaning-up operation to come.

"You're lucky really," Dug said with a grin, "in the other corner is a can of creosote with the cap off…"

I could train a dog, handle horses and cattle, but why, I wondered, could I not control two small boys, my own offspring?

Then Evelyn came to my rescue. She was a school leaver from a neighbouring village who wanted a job with children. The eldest of a large family with an ailing mother, she was well used to the care of children. She was tall and thin, with straight, dark hair framing her pale face – as different from blonde, buxom and pink-cheeked Mary as could be imagined – but her frail looks and quiet manner belied a determined disposition. From the day she arrived in early August, she was in control of the boys. Moreover, she liked washing and ironing.

As a further compensation, when at last I could get back to my typewriter, I sold an article and a short story based on the boys' escapades.

5. Mainly Small

Of course there were the occasional intrusions of the larger farm animals, such as a herd of cows on the front lawn when the drive gates were left open, a common human error.

Then there was the afternoon I came in from a walk with the boys to find not only the gates but also the front door wide open and a rumpus going on in the dining room.

Four young piglets were playing tag round the table.

There was an element of surprise on both sides and at the sight of me they bolted from the room just as the boys appeared in the doorway. There was a melee of arms and legs both human and porcine and mingled cries and squeals, then the pigs rushed away down the drive, straight into the arms, or rather legs of Sam, the boys still wondering what had hit them.

Most of the animals that joined our household did so by invitation, or at least by capitulation on my part. There were always cats. They turned up as if from nowhere then disappeared just as suddenly after the manner of farm cats. Usually they stayed around the buildings, only coming to the house to be fed.

The first of these 'outside cats' to become an 'inside cat' was Freda. The boys found her in the yard one morning, sleekly black and white and very pretty, but so nervous that they immediately dubbed her 'Fraidy Cat'. Their efforts to tame her were so successful that soon 'Fraidy' was inapplicable and became Freda. From there on she took up a place by the Rayburn by day and sat on Peter's lap by the fire in the evening. Like all female cats she produced kittens with unfailing regularity, each litter being received with great enthusiasm by the boys.

Freda's pattern of behaviour on these occasions never varied. She would have her young in the straw in the Dutch barn, but when their eyes were open she decided it was time for me to take a hand in their rearing and literally dumped them on my doorstep.

The first time the operation was carried out in easy stages. One morning of teaming rain I saw her bringing kittens from the barn to a clump of nettles in the corner of the flint-walled farmyard. As the morning wore on with no abatement of the deluge, I became anxious about them in a place of so little shelter. Then Freda's face appeared at the kitchen window, her wet fur standing out in exclamation marks round her face, her mouth wide open in an anguished but silent 'meow'.

I went to the door. On the mat was a very wet kitten. By the time I had found a suitable box, filled it with soft shavings and placed the kitten in it by the Rayburn, Freda was again at the window and on the mat, a second soggy kitten. She brought two more, then settled down happily with them and began to lick them dry.

The nomenclature of kittens occupied the boys thoughts for some days after each litter was born. It helped, they said, to use the same letter of the alphabet, so we had one family of attractive greys and tabbies called, Minx, Moley, Mousie and Minnie. Pampered because of their pretty faces, these kittens became very tame; too tame. Give a cat an inch and it will take a mile. These four took over the kitchen and I was soon at my wits end to keep them out of the cupboards, off the tables and their heads out of the milk jug and anything else they could get into. They slept in Jane's basket and although she did not exactly welcome her bed-mates she was too long-suffering to take any action against them.

If I put them out of the door they re-entered by the window until at last I put up a netting to keep them out. Denied entrance they wreaked vengeance on the garden, to the detriment of vegetables and the herbaceous border.

Eventually they became civilized. Minx was devoted to me and frequently brought me gifts, such as a dead rat placed thoughtfully in my chair or a live field mouse at my feet. Moley, never quite sure which side of the door she wanted to be on, became known as the 'boat race cat' because she was 'in, out, in, out'.

Mousie, when time came round full circle, distinguished herself by having kittens in an armchair. Such a thing had occurred before, but on this occasion, I was sitting in the chair when the happy event took place. She jumped over the arm and snuggled down behind me as I sat in front of the fire. As the arrangement was warm and comfortable for both of us, I let her stay. Presently I felt some curious movement at my back and looking round I discovered not one cat but three.

Peter, always quick to act, fetched what we now called 'the emergency box' and gently transferred Mousie and kittens, now one more, into it. Apart from a baleful glance, she made no objection, and when carried out to the stable gave birth to two more.

When next she became pregnant it was a battle of wits between us to prevent her having her family in the house. I won that round and they were born in the Dutch barn.

A week later I was working in the front garden when I saw my feline friend trotting briskly up the drive with a kitten in her mouth and a purposeful look in her eye.

She disappeared through the front door and with thick mud on my shoes I was unable to go after her. A few minutes later she came out and shortly returned with a second kitten.

This time I was ready for her. I followed as she shot up the stairs, but before I reached the top she had disappeared like a puff of grey smoke.

The same thing happened on her next two trips.

I diligently searched the three top rooms and the box room but there was no sign of Mousie or the kittens. During the next few days I frequently met her on the stairs but always she disappeared as if by magic, so the boys and I evolved a plan.

When next we saw her coming they would run up and hide, one each side of the top landing and watch where she went. It was Peter, in the west side room who saw her hurry in and make for the box room, where against a cupboard there was a piece of floorboard missing. She slipped underneath it and there under the floor, we discovered her four kittens. It was another case for the emergency box and the garage. She never returned to the house.

It was inevitable that in the spring the boys' should adopt a motherless lamb. They took turns at feeding it from a bottle and they must have done a good job for Thomas became strong and very playful. He responded to Jane's efforts at mothering by chasing her round the lawn until she ran, disconcerted, into the house. He would have chased Mousie had she not surprisingly, stood her ground. The two afterward became fast friends.

When he grew older we tethered him on the lawn to eat the grass but save the borders, for like Ferdinand the bull, he not only loved to sit and smell the flowers, but ate them as well. After the day I returned from Chichester to find that he had broken his rope and stripped the rose bed and herbaceous border of blossoms, I never felt quite the same towards him.

One spring Dug was asked to supply an orphan lamb to an old lady who lived in one of the farm cottages. She was lonely, she said, and it would be company and crop the grass, which she could no longer cut. Dug selected a nice tup for her and she was delighted. She named him Alfred after her late husband, and very soon like Mary's lamb, it followed wherever she went.

No one remembered that in the transaction Alfred had escaped being castrated with the other young males. In time he grew into a very fine ram. The old lady was proud of her handsome pet and they were very happy together. Alfred's manners both indoors and out were irreproachable until one day his mistress received the present of an antique mirror. While she decided on a permanent place for it, she put it on the floor of the landing. Next morning, Alfred, trotting upstairs as usual, came face to face with another ram – a rival.

For a moment he stared at the intruder, then, jealous of his territorial rights, he charged... There was a crash and a shower of splintered glass and Alfred stood, dazed and bewildered, an empty oak frame about his shoulders.

The old lady, of course, forgave him.

"T'was only natural," she said. "Rams do allus fight. Wot does I want wi' a mirror, anyway. I lost all me beauty long since."

One Christmas the boys were given three guinea pigs, charming little sleek, sable and white creatures which they named Gold, Frankincense and Myrrh. Alas, they were short-lived. On uncovering their hutch one morning I found that a rat had got in to them in the night. The result was a macabre scene and I hastily removed all evidence of it before the boys came out.

After we had reinforced the hutch with smaller mesh netting, they were replaced with three more which became a source of pleasure and amusement for a long time. They received the same names as their unfortunate predecessors.

The boys loved to carry them about, tucked warmly in cardigan or jersey. Many times a day I was called to the rescue of Gold who had become wedged in the armhole of Nigel's blazer or Frankie in Nick's sleeve or Myrrh who was halfway up Peter's back.

They played with them in the sitting room in the evenings but after they had gone to bed it was left to Evelyn, to hunt for three elusive guinea pigs under the sofa or bookcase and carry them back to their hutch.

As they turned out to be one male and two females and the latter produced a family every six weeks, we soon had a small herd and quite regularly supplied a pet shop in Bognor.

After a while the boys tired of them, and the cleaning out and feeding of sixteen guinea pigs was left to me. Eventually I too became tired and the pet shop obligingly relieved me of them all.

No sooner had they gone than we went to stay with friends who had two young boys. The boys had pet mice. Like the guinea pigs they were sable and white and very engaging. It was a foregone conclusion that we should bring some home. Like the guinea pigs they multiplied rapidly and were accommodated in cages in a loft over the well house, but were frequently brought into the house.

If catching guinea pigs under a bookcase was a tricky business, mice were worse. At one time we lost them down a hole behind the television set where they spent two weeks cohabiting with the house mice. After they reappeared and were caught I anticipated some strange markings in the next litters of young, but they remained true to type.

House mice were always with us and if messy and destructive so that I was sometimes reluctantly forced to set traps, they too provided us with entertainment. One night a loud tapping on the floorboards awakened us. I put on the light and was amazed to see a mouse trying to get into its hole, but prevented from doing so by a dog biscuit stuck on its front tooth. It ran frantically up and down, banging the oval on the floor in an attempt to dislodge it. Seeing us, it made again for the hole and this time it freed itself from its impediment and disappeared.

Beside the common house mouse we had visits from outdoor mice just passing through. There was a long eared field mouse which appeared each morning through a hole in the ceiling, just above the cooker, where the pipe entered. Sliding down it like a fireman, he would sit on the back of the stove watching me cook breakfast. Afterwards he would mount the rim of the cooling pan, eating what scraps he could find. He became a pet and a source of great amusement to the boys. I dared not set a trap during his stay.

Once, to my great distress, I caught a beautiful yellow-necked mouse in a trap I had set for his commoner cousins. This rare mouse is larger than the rest of the species, reddish in colour with a yellow neck, large eyes and incredibly long whiskers and tail.

No menagerie is complete without a tortoise. Freddie was a handsome fellow with a dark, beautifully marked shell. Dug made a wire run for him on the front lawn or he was allowed to roam the walls of the garden under supervision. The boys fed him on lettuce, spinach and occasional bread and milk, but he most enjoyed a fat, golden dandelion flower and it was amusing to see one gradually disappearing into his small mouth.

Then came the disastrous day during harvest when Freddie was reported missing, having burrowed out of his pen. A search went on all day, along the bank, in the borders, the adjoining field of stubble; no bush, plant or blade of grass was left unturned, but no sign of a tortoise.

A week passed and hope had practically died when John came to the door after dinner, holding out Freddie in the palm of his hand.

"Found him on the platform of the baler," he explained. "Lucky thing I didn't start up, he could have ended up in a bale of straw!"

"Then we might never have found him," said Nick.

The situation had all sorts of possibilities. I used some of them in my first short story for children, called 'Freddie's Adventures'.

We kept him after that for several years, putting him away in a box of straw in the cellar each winter until one false spring in February, when he must have woken too early and in the subsequent cold weather, with April snow, he died.

One morning Dug called me from the kitchen.

"Come and see Will's ferret!"

Will was in the yard with a lump on his shoulder, the ferret having at that moment decided to explore underneath his jacket.

"Cum out, Bessie an' show yersel'," he wheedled. A sharp white face peered from under his lapel, then a long creamy white body oozed across his waistcoat. "This 'ere's Bessie. She's a good ferret an' as 'armless as a kitten." He allowed her to climb up his sleeve and round his neck to demonstrate her harmlessness. "Quiet as a kitten."

Bessie fastened her fierce crimson eyes on me and I didn't think she looked all that quiet, but nevertheless stroked her when invited to do so. Her fur was deep and soft. She was undoubtedly beautiful.

"As 'armless as a kitten," Will reiterated. "There's a rabbit bury in t'paddock, shall we try un?"

We followed Will to the paddock, he looked at me doubtfully as if unsure whether a woman should be allowed on such an operation, but I was determined to see Bessie in action.

Will drew a net from his capacious pocket and having popped Bessie down the hole, spread it over the top. I was told to keep absolutely still and quiet. Very soon there was a loud rumbling from underground.

"Aah, 'ee be at 'ome," said Will with satisfaction.

It was incredible to me that two small animals scampering underground could sound like a train in a tunnel, but the rumbling went on, even louder.

Recklessly I moved a foot and a twig snapped like a pistol shot. Will glared. There was silence after that for a while, then more rumbling. I was seized with the desire to sneeze but dared not.

At last a rabbit appeared at the hole. We held our breath, but it slipped the net and was gone.

Will pulled the ferret out.

"Come on Bessie," he said. "'Tis no use tryin t' ketch a rabbit wi women around."

I never could see just how I was to blame.

I have never liked to see birds in cages. Blue tits feeding on the nuts outside the window give me far more pleasure than more exotic species behind bars, but when the boys wanted a budgerigar, I gave in.

Mickie was quite an ordinary green budgerigar, he didn't talk or do acrobatics and, as I anticipated, the job of feeding him and cleaning his cage eventually fell to me, but he was lively and companionable and lived happily with us for several years.

At last the boys decided that he ought to have a mate and we bought a little green hen. Mickie was estatic with joy and she reciprocated his affection. Their matrimonial bliss was touching to behold and we felt rather guilty that we had deprived him of feminine company for so long.

Then, before many weeks had passed, tragedy struck. She became ill with some virulent disease and before we could isolate her, in her paralytic fluttering she struck him down, breaking he neck. She to fell dead shortly afterwards and they lay at the bottom of the cage like two ill-starred lovers in a Shakespearian tragedy.

To my relief there were no more caged birds, but in consolation for the loss of the budgies, the boys were given a pair of white fantail pigeons. I was delighted. Fantails were the one thing needed to complete the old worldliness of our walled garden. There was something about them that made me feel like getting into a long dress, floppy hat and lace mittens.

Fantails, we found, competed very nicely with our former pets over the matter of multiplication and in no time the garden was a mass of white birds, which seemed to spend their time eating and mating. From every window I could see billing, cooing, necking, copulating pigeons; on the stable roof, the henhouse, the bird table, all the upstairs windowsills and on top of the cistern in the loo.

When we entered the garden, they rose in a white cloud; a line of pigeons stretched from end to end of the stable roof; they ate the hens' food and made raids upon the kale, (tactfully they did not touch the garden vegetables like the wood pigeons). They made an appalling mess around the buildings and on top of the car. Dug said they would have to go and eventually, they did, but no sooner had we seen the last of them when Nick arrived home from school with a basket. In it were two homing pigeons.

"Oh, Nick dear," I exclaimed. "What's the use? It will be the same thing all over again."

"No it won't," he replied confidently. "We'll build a proper nesting place and when they get too many, we'll collect the eggs."

He made it sound quite simple but I did not feel so sure. I wondered what Dug would say.

With unexpected good fortune for Nick, it turned out that Dug had kept racing pigeons way back in his youth in Wales and was full of enthusiasm, reminiscences and advice. He helped to build the nesting place in part of a new shed I'd recently got put up for some pullets. I wondered as they worked happily away how much room I was going to have left for *my* birds.

The pigeons approved their quarters and settled in. The cock was rich reddish brown and the hen a pretty blue pied. Their young should be interesting. Pigeons mate for life and are extremely faithful. The cock takes equal turns with the hen at incubating the eggs and I could have timed my cooking by the punctuality with which this pair changed shifts.

They led a busy life, each pair laying two eggs every month. As soon as the young had learned to fly they started over again with another two. This time we were determined not to let things get out of hand, so when there were eight birds we started to collect the eggs.

When hens eggs were short, I used them in cakes and puddings. They rose beautifully. In time the boys decided to train the pigeons. Taking the oldest pair they took them several times up to the top Down and released them. What route home those birds took we never knew, but each time the boys arrived back hours before they did.

They tried again with a younger pair with no better result and finally gave up, so my bright idea that Dug should take one in the Land Rover and release it with a note when he was going to be late for dinner never reached fruition.

Our own pigeons attracted others passing over and we often had a stranger with us resting for a few days. One morning Peter found a handsome dark blue bird on the ground by the shed, completely exhausted and pitifully thin, too spent to take either food or water.

We brought him in, put him in a box and fed him at frequent intervals with milk and glucose from a nasal dropper. After a few days he was able to take a little brown bread in milk and then corn.

He stayed with us for six weeks, at first just strolling about the garden then taking short flights, then flying over the fields taking his exercise in daily widening circles. One morning he circled twice and was gone.

There is no egg as delicious as a bantam's. For years we kept some of these attractive little hens in the orchard, starting with Mrs Black Stocking and Mrs White Stocking, so named because of the contrasting hues of their legs and Rastus, a bold and colourful cockerel who made up in audacity and daring what he lacked in size. In time we had a number of bantams. They came in various shades from speckled brown to cream and a pretty shade of apricot. Excellent flyers, they came over the wall and did considerable damage in the garden, for which we forgave them because of their engaging personalities and their unfailing supply of eggs.

Collecting their eggs resembled a treasure hunt, for they laid in some strange places; one in an old car, another on a tractor seat or a disused pig sty. In summer they laid out and we failed to find all the eggs. Then a hen would disappear for a few weeks, emerging from a clump of nettles, followed by a brood of chicks.

A few days after Mrs Black Stocking had appeared with a family, Peter ran in to tell me that disaster had befallen them. I ran to see. In the yard, a drum of molasses had leaked and six tiny chicks were floundering up to their necks in the black sticky stream while Mrs BS jumped up and down squawking in consternation. It took a lot of washing in warm soapy water and drying by the fire to restore them to their former yellow fluffiness. The adventures of the Bantams provided me with material for several children's stories.

For a while we had a pair of Aylesbury ducks in the orchard. I like ducks. They are more intelligent and companionable than hens, but they have one irritating characteristic. While a hen will find its perch at dusk, a duck wanders to and fro outside its house, waiting to be put to bed. I remember that filling their zinc pond entailed much carrying of water in dry weather and that they were fond of lettuces, disposing of all our bolted plants. I forget what became of them.

Beside the more orthodox pets, I was sometimes brought young creatures of the wild to raise. I am not generally in favour of removing from its environment a supposedly abandoned or orphaned young animal, for usually the mother is not far away and will come back for it; this especially applies in the case of fawns. However, when Nick brought in a young tawny owl, I could not make him take it back.

It should be fed, he told me, on a fresh caught mouse every day. First catch your mouse! Live mice we still had in plenty, but freshly-caught ones were harder to come by. However, I felt that Jane and Beauty not begrudge him a few small pieces of their daily allowance of raw meat and he did very well on this, fed to him with a pair of tweezers. He also required grit and feathers to form the pellets by which he ejected waste matter.

As Nick was at school all day I attended to these needs, but Tawny failed to show a fitting gratitude. While he would sit on Nick's shoulder or hand in the friendliest manner, he always greeted me with a loud, prolonged hiss, which apparently, in owls is a sign of intense dislike. As he grew stronger we let him out for exercise. At first he would fly the length of the lawn and back to Nick's shoulder, but gradually he went further afield, until finally he took up residence in a beech tree by the farm buildings, where we used to see him for many months afterwards.

I was not so successful with my next charge, a baby dormouse no more than an inch long with auburn coat, minute feet and a splendid set of whiskers. After a week he was taking milk and glucose so well from nasal dropper that I got carried away with my task and overfed him. He passed away in my hand, suddenly and quietly, in the middle of a feed. The boys received my apologies with an air of frigidity and later, when they were discussing whether or not they should adopt some other creature, I overheard the remark, "Oh well, if we want to get rid of it we can always give it to Mummy to feed."

This I thought was a little unfair. After all, I had never set out to be Jacquie Durrell.

In addition to routine feeding and care of animals I was sometimes called upon to carry out autopsies on suddenly deceased guinea pigs or rabbits (oh yes... we had a rabbit or two) or to set the broken wing or leg of a bird.

I only once attempted surgery. My victim (and I do mean victim) was a hen. Chicken should never be allowed to get at long loose grass for if they swallow it whole it winds round in the crop, forming a hard ball. This was the problem with my hen.

"It is quite easy to operate on her and take it out," said our farm secretary, who had long experience with poultry. "You just open up the throat, then the crop and take out the grass."

It sounded simple. The next day, with Evelyn as a reluctant theatre sister, I set about the job, having carefully sterilized a razor blade, needle and thread and swabs. Hens are such silly things. Although you feed and care for them every day and they should know you well, they never allow themselves to be picked up without hysterical shrieks of "Help! Murder! Police!" accompanied by a frantic struggle.

Having got this over, my patient settled down quite calmly while I sat on a low stool with her head held downward between my knees. Carefully, I made my first incision, found the crop, which was packed hard, cut again and pulled out the evil smelling grass. Then I swabbed out the crop, feeling rather pleased with myself. So far, so good.

Then my troubles began, I had made two incisions as there had been two separate skins, but now they had merged into one, and try as I might, I could not separate them.

My patient was getting restless and her struggles increased my difficulties, I visualized her flying away with open throat and crop.

In sudden panic I sewed up the one skin and released her. For a few days all seemed well; she was eating and chortling away to the other hens, obviously telling them all about her operation. Then an ominous swelling appeared in her throat. The next day she looked poorly and had a very peculiar smell. The only thing was to dispatch her with all speed.

"You should have made a much longer incision in the neck than you did in the crop," our secretary told me. "Then the skins would not have joined together. Never mind, you will know another time."

There wouldn't be another time; in future the only chicken I intended to carve into would be dead and on a dish.

6. Come Down and Play

A village expects that every man shall do his duty. This entails contributing to the activities of the local branches of various organisations; the Womens' Institute, the Mothers' Union, Red Cross, Youth Clubs and of course the Church – holding or assisting with stalls at fetes and jumble sales, arranging the altar flowers and serving on committees.

By the second year at Court Hill I had become embroiled in a number of these activities but not to the same extent as Dug, whose chief commitment was that of Rector's Warden. His duties were many and included leading the Bishop on ceremonial occasions, entertaining visiting clergy and coping with the church drains.

He also had to be in church half-an-hour before the rest of the congregation, during which time I sat in idle solitude in our pew. Often I passed this time, not perhaps as I should, with some uplifting reflection, but with weaving stories around a certain Tudor Squire whose effigy lay in the south-east corner of the nave.

The church was very proud of its 'knight'. Carved in oak, his was the only wooden effigy in Sussex. The recumbent figure was clad in plate armour of the late Wars of the Roses; his hair was long, his bare head resting on his helmet.

He was Anthony St Leger of Binsted near Arundel, who died in 1539, requesting in his will to be buried in the church of St Mary in Slindon. The blackness of the oak and the deep fissure down one side of his handsome face, the wearing away of one foot, testified to the long centuries he had lain there, his head turned towards the light that slanted from the south window.

Why, I wondered, was there no name or dedication on his tomb and why should he wish to be buried at Slindon instead of in his own lovely ancient little church at Binsted?

Was it a romantic reason? Had he returned from a battle between York and Lancaster sickened with a wound that would not heal? Could he have met a Slindon maiden who cured him with herbs from the walled garden of the Manor and did she lie also in Slindon churchyard?

There was an aura of mystery and romance about this nameless warrior that had fascinated a youthful Hilaire Belloc and inspired him to write his first poem:

> *There is no name upon his grave*
> *If his grave it haps to be*
> *And his face doth look towards the plain*
> *And towards the calm blue sea.*

Here was a subject for a historical romance, I thought. Perhaps, if ever the day came when I had plenty of time…

To Dug's role of Rector's Warden there was added that off Chairman of the Parish Council, the Conservation Committee, the Village Hall Committee and President of the local Cricket Club. His life was often a round of meetings. Finally he was appointed foundation manager of the village school. This he thoroughly enjoyed for it enabled him to get among and to know the children and his popularity was obvious. Little girls especially leaned over the playground wall to wave

71

to him when he drove by in the Land Rover and ran to meet him when he entered the playground. It was the girls who neatly penned the invitations to the managers' luncheon or their Christmas party.

He soon knew every child by name, but one day on one of his visits to the classrooms, noticed a strange blonde head. Its owner was seated on a low stool and was bowed over a book.

"And who are you, my dear," Dug asked, a little glutinously, smoothing the fair hair.

The girl looked up and eyed him coldy.

"I'm the student teacher from Bognor College," she replied with dignity. It took some time to convince the headmaster that it was a genuine mistake.

As each of the boys attained the age of nine, they joined the church choir. At times they were its only members and became known as 'the family choir'. Gone were the days when the men of the village and the boys of the church school swelled the ranks of the choir until the singing on a Sunday morning could be heard all over Slindon.

Their services were also in demand for weddings. It was always a difficult decision on a beautiful summer Saturday afternoon – to pass up the then munificent sum of two shillings paid to each choirboy or go to the beach.

In 1956 Slindon House was taken over by a boys' school and the choir was much enlarged in consequence. From having the vestry to themselves, our trio now joined in the general scrimmage for cassocks and surplices, a problem Nick solved by climbing to the top of a tall step-ladder to change into his.

By the time all three boys were at day school, Court Hill had become, and continued to be for many years, a centre for all the youth of the village. Anyone at a loose end invariably found their way down to the farm, where there was sure to be some activity.

Firstly there were the boys from the farm, Ken, the pig-man's son and David, known as Di, and Ray, the son of our herdsman. John, whose father was the local policeman then two boys whose parents kept the beer shop and small general store at the top of the village, joined them. Richard, who was called Pee-wee and David who was Tom. There were a few girls too. Eunice and Irene, daughters of the family grocer and Hazel and Vivienne who came along with their brother Lance. They were the children of an Army Captain domiciled in the village. Then there were Henry and Charlotte, whose father was a Squadron Leader and lastly Nigel's special pal John, who lived with his grandmother in the village and because of his height and a certain rotundity of figure was known as 'Big John'.

Besides these 'regulars' were a varying number of hangers on, so the total gang numbered in the region of twenty.

There was never any recognised leader, although Nick was usually the instigator of their various escapades. Often it was Tom, a tall, lively good-looking boy with a mop of brown curls and mischievous blue eyes who took the initiative. Sometimes the others looked to Di, who was a little older, with a grave manner and a quiet smile, which suggested a greater knowledge and wisdom.

An almost continuous activity during the summer holidays was cricket, played on the front lawn. As soon as one lot of players dispersed to seek another amusement, more would take their place. The tackle was kept in the summerhouse and even if we had been out for the day we would invariably find a game in progress on our return.

Ken was a fast and furious bowler, responsible for putting more than one ball through the front windows. There was an awful moment one afternoon when I was entertaining to tea, certain ladies from the village. We were talking quietly when there was a shattering explosion and a shower of broken glass. A cricket ball whizzed past my ear and landed in the lap of the Rector's wife. Luckily she was more frightened than hurt but my tea party broke up very quickly afterwards with polite but shocked farewells. Ken was banned from bowling for some time.

A rival pastime was the construction of camps in the wood. When a good camp was in progress, the boys would disappear after breakfast, accompanied by Jane and Beauty, who were accepted camp members, and only reappear at mealtimes.

From time to time I was invited to come and inspect the campsite and the building. This was a carefully constructed frame of all the loose wood that could be gathered, roofed with branches of conifer, sometimes reinforced with an old tarpaulin. Usually, if there was sufficient wood, a corral was built outside.

Furnishing consisted of old chairs, milking stools, orange boxes and any kitchen utensils and crockery that could be scrounged. Outside the camp hung a long rope from which the members of the gang could swing from place to place. It became no surprise to me when walking in the woods to see a boy swinging high, like Tarzan among the trees.

The site was always carefully concealed and wrapped about with an air of secrecy. No-one outside the gang, except myself, was supposed to know its whereabouts. If discovered, it was hastily dismantled and the business of building on another site began all over again.

At times when the gang became too large, it split into two. This resulted in rival camps and mock battles on the Downs, but when a summer school came to Slindon House, a deadly feud arose between its boys and girls and the Court Hill gang. Now the battles were in earnest. One day I was walking peacefully home with the dogs toward North Wood when Nick, Peter and another of our gang came running along the road with ten or twelve of the school in hot pursuit. Our boys were looking anxiously over their shoulders and appeared to be really frightened.

"Help, we're being attacked", they cried when they saw me.

I could see that this was no joke, but it seemed unlikely that I could offer much protection. The school gang were no doubt enraged by some prank played upon them by the others and were after their blood.

Both they and I reached the North Wood buildings together and to my relief I saw Sam in the calf pens. Sam was a good person to have around in an emergency. He was noted for his presence of mind and prompt action. Hearing the din he ran out, armed with a pitchfork. The school gang checked at the sight of him and our boys quickly took refuge behind him.

There we all stood, Sam with his pitchfork holding the enemy at bay like a small town sheriff holding back a lynch mob and telling them to go home. I added stoutly that if they did not I would set the dogs on them. They rather spoiled the effect of this by wagging their tails while Jane put on her most disarming grin. I hoped they would think it was a snarl.

After a good deal of muttering they decided to turn back and the boys walked home with me. Sam said he had to go to the farm on the tractor anyway so he came along as escort.

"What did you do to make them so mad?" I asked as we went.

"Oh, we kidded them that there were secret passages in the grounds of Slindon House," said Nick.

"Well, there are supposed to be," I replied.

"I know, but we showed them a place in the wood where we said one came out and they could get through to the house cellar and they were digging all morning."

"Old Will saw them," added Peter, "and told them that no one knew where the passages were."

"Serves you right that they came after you," I laughed.

We heard no more about the incident.

The summer holidays came to an end and the evenings drew in. The gang could only meet at weekends. Now football took the place of cricket on the lawn. I thought that a football was less likely to come through the windows. I was wrong. Instead of camps in the wood there was a clubroom at the top of the house, furnished with a screen and small tables and chairs and illuminated by candles in bottles. Here they played poker in an atmosphere so redolent of a TV western that I could almost hear Doc Holliday's cough.

I had been told that Hilaire Belloc chose this little room as his study and I wondered what he would have thought of its present use.

When they reached the age of ten the boys went beating for the shoot. Instead of a Saturday morning lie-in they had to be prised out of bed in time for a substantial breakfast to sustain them through a long, cold or wet morning in the kale or the covers and to be ready to start beating in at 8.30 am.

There was the inevitable scramble over putting on the layers of clothing to protect them against the elements. Over-socks, gloves and leggings chose this worst possible moment to play hide and seek. Peter's woollen hat would mysteriously turn up in someone else's pocket while Nigel's beating stick was nowhere to be found.

Arguments ensued until Dug's voice, loud and irate, rose above them, demanding to know why they could never find anything, that all this should have been made ready the night before and had anyone seen his cartridge case?

At last they were ready and straggled out one by one to join their pals and the senior beaters on a trailer waiting by the cart shed. Beating was not only a source of pocket money and good exercise but also an opportunity for observing the wild-life of the woods. Deer, foxes, hares and rabbits were numerous and many birds that did not come to the garden.

That winter Peter joined them. The three pooled their season's beating money and bought a second-hand record player. It was a good one and their especial pride and joy. They kindly said that Dug and I might use it, which we would have done, had we been able. It was a funny thing. Although we had both operated a variety of machinery, both agricultural and domestic, the technique of setting a record in motion on the turntable was completely beyond us. It was automatic too. All one was supposed to do was to touch a switch and the needle arm would rise from its rest, move to the left and lower itself gently onto the record ... then music.

But when we touched the switch, the arm rose and after weaving to and fro for a bit like a charmed snake, returned to its rest and the machine switched off.

One Sunday morning Dug came in to breakfast before the boys were up and as the record player was in the kitchen it seemed a good notion to take advantage of this unusual peace and solitude to play for him the Harry Secombe record I had bought for his birthday. I put the record on the turntable and touched the switch. The needle rose and hovered while we waited for the dulcet tones of my second-favourite Welshman. Then it went into its snake routine and returned to the rest. I tried again and again while the toast burned and the milk for the coffee boiled over.

I called Nick, who came down in his pyjamas, shuffling across the kitchen like a sleepwalker, flicked the switch without seeming to open his eyes and went back to bed.

We munched happily on the burnt toast and overdone eggs while the mellifluous Secombe tenor filled the kitchen.

When the last note died there was silence for a moment, Dug and I looked at each other.

"Shall we have the other side?" he asked diffidently.

"Why not," I said.

This time it was Peter, a little more awake and half dressed, who came to our rescue.

"I can't think it why you can't do it," he remarked with a superior air. "It's so simple."

That must have been the reason. It was too simple.

In the spring Dug bought two air guns, which the boys used under supervision. The practice range was in the back garden; a target nailed to the cherry tree, a row of tins on top of the wall and for the crack shots, bottle tops suspended on strings, to be fired at from the top of the garden steps, a distance of about twenty yards. It usually fell to me to supply the ammunition when I went shopping and I turned up so often in the gun department of a certain Chichester sports shop they must have thought I was Annie Oakley.

A tin of five hundred lead pellets added considerably to the weight of my shopping basket, but in return I was allowed to take my place on the range. I never hit the bottle tops but I scored pretty well on the target and the tin cans.

About this time we had another influx of rats and every day seven young played among the sweet pea sticks in the garden.

I had the .177 trained on them from the kitchen window and the .22 from the bathroom, but they were elusive among the sticks and I never hit one although I could get the sweet pea packets dead centre every time.

I was not so successful as Maid Marion when archery became the favoured sport. My arrows usually fell short of the target pinned to a bale of straw by the apple trees at the bottom of the front garden. Now my trips to the sports shop were to replenish arrows lost in a cornfield or in the branches of sycamores.

I had a few hobbies of my own. One was ginger beer making, which in time became not so much a hobby as an endless chore. There must be very few housewives who, having happily accepted the gift of a ginger beer plant, have not eventually wished they had never seen it. It became a taskmaster, a tyrant. The plant had to be fed with sugar every day for six days. On the seventh day, come hell or high water or however many full bottles already stacked the larder shelves, another eight pints must be made.

Even with the help of the gang, our drinking capacity could not keep pace with production, neither could we give it away except to unload a few bottles on a stall at one of the local fetes. Then the problem arose of finding enough bottles. All that we could find, beg or borrow were soon filled.

"There are plenty of empty bottles in the church vestry," said Nick one day when I was wondering what to do with eight pints of loose ginger beer. "I don't think anybody wants them."

The next Sunday he asked the Rector, who said he might have them. We fetched several crates, and for a while I had plenty of bottles, labelled 'Communion Wine'.

In the end I gave away my plant to another unsuspecting housewife.

Sometimes on a Saturday afternoon Dug would take the boys fishing in the millstream at Arundel. This deep stream of clear pure water flowed from the watermill and under a bridge, both of which replaced others that had stood from earliest times. I have since learned that fishing is not allowed in the millstream so I hope this disclosure will not lead to conviction! I tagged along on these expeditions but my recollections are mainly of slithering up and down the greasy bank, doling out revolting-looking bait to which the large shoals of dace seemed, not surprisingly, completely indifferent; getting my hook caught fast in a tree root on the only occasion I was invited to cast a line; rescuing Nick who was prone to falling in, and nearly sharing this mishap when trying to reach a piece of purple loosestrife or a giant kingcup for someone's wild flower collection.

Sometimes they caught a few less wary dace and placed them in a keep net, to be released at the end of the afternoon's sport, or they got charming little sticklebacks in a net and brought them home to swim with the goldfish in the garden pond.

Tired of fishing we would walk along the narrow path to where the stream, almost jade in colour in its deepest parts, flowed over a small weir into the muddier waters of the Arun. Here tall yellow iris bloomed by the water's edge and the alders boasted their white feathery blossom. Swans and mallards swam on the river while coots and moorhens darted in and out of the reeds and bulrushes. Sometimes we saw a flock of cormorants flying in formation down river from Amberley Wild Brooks to the sea.

"Isn't it a beautiful scene?" I would remark to Dug.

Inevitably his reply would come.

"Hummph, you should see the Welsh rivers!"

Ah yes, the Wye, rippling crystal clear over stones and boulders, the silver gleam of a leaping fish, the blue flash of a kingfisher on the wing, a dipper poised on a rock before his dive; but I am always content with the beauty I find at hand.

On the other side of the bridge was Swanbourn Lake, which fed the mill stream situated in a corner of Arundel Park. It was banked on one side by the beech-wooded ascent to the castle and on the other by undulating slopes and knolls of parkland. Swans, ducks, coots and moorhens nested on an island and around its banks and in spring cynets and many broods of dark or speckled chicks clustered in the reeds.

In summer we could take a boat and row on the lake.

"Let's row over the bomb hole," the boys would always urge. The hole was deep and wide under the hanger wood, the water black and impenetrable and rather sinister. I was always glad to see again the weeds on the bottom. When autumn came the lake reached the climax of its beauty with the varying hues of gold and bronze and russet of the beech woods reflecting in the water. Ducks were there to be fed at all times of the year but in winter they would march off the pond like an advancing army as soon as any visitor appeared with a paper bag.

The road curved on past the lake, beyond the Black Rabbit, an early 19th century inn, up a steep hill between chalk pits on the left and spreading marshes to the right, to the little hamlet of South Stoke. It consisted of a farmhouse and few cottage and a tiny 12th century church.

As we descended again the narrow winding lane, the ex-panse of marshes spread out before us, with the meandering

Arun, the little brooks and the dotting of black cattle. Dominating the scene was the castle on its hilltop.

I was about to exclaim on its picturesque splendour, but remembered there are also castles in Wales...

7. Horses and Dogs

The woods and Downs were a popular area for horsemen of all ages. Riders exercising hunters frequently passed the farm in the early morning. In the woods we met strings of ponies with young riders or found evidence of their presence in the improvised jumps of brushwood obstructing the paths.

Not infrequently we spent an afternoon at weekends catching a riderless horse and looking for its owner or assisting a horseless rider to find her mount.

It was some years since Dug or I had been in saddle and we began to think that it would be nice to have a horse again. Then we received an unexpected letter from an old friend who had farmed a small acreage near to where we had lived in Hampshire. Mrs Chalcraft bred pedigree Dexter cattle, kept several Springer spaniels and drove a 1923 Roll Royce coupe. Often we saw her bowling along the lanes, her tall, spare frame erect behind the wheel, an aged spaniel sitting just as straight beside her.

Dug often went to see her to give advice on her farming. Sometimes I went with him and over tea in her lounge, adorned with paintings and photographs of horses, cattle and dogs and silver cups won by all three, she talked about the days when, an enthusiastic rider to hounds, she had kept twelve horses in her stables many of which she had bred herself.

Although her riding days were over, she still had a few old retainers lent out to various friends of the chase. It was about one of these horses that she now wrote to us.

Stoburgh – Stobie for short – was a sixteen-year-old bay gelding by Newburgh, a pedigree racehorse out of a Canadian mare named Queen's Birthday. He was now too old for hunting but there were useful years in him for hacking. Would we like him? Indeed we would.

A further letter told us that he was at a livery stable in Bishop's Waltham and we could collect him from there, complete with saddle and bridle. Enclosed was a photograph of Stobie in his heyday, being ridden side-saddle by Mrs Chalcraft at a show where he had won a first in the heavy hunter class.

Stobie was inclined to be headstrong, she warned us, and having been reared and ridden by herself and girl grooms, he was suspicious of men, but she was sure that Dug and I would handle him. A week later we set off for Hampshire in the cab of a horsebox with our local haulier. It was a perfect April evening. Cottage gardens were aglow with flowering trees and fragrant with stocks and wallflowers. On the outskirts of the New Forest dainty silver birches wore a delicate, green filigree and on the common the gorse flamed saffron in the westerly sunlight above new fronds of bracken. We had no difficulty in finding the livery stables. A stocky man with a florid face and hard blue eyes came out to us. His greeting threw cold water on our high spirits.

"You've come for the big hunter. I'll be right glad to see him go, real devil he is. We don't get many like him, thank goodness, nearly killed my man yesterday."

Dug and I stared at each other.

"There must be some mistake," I whispered. "Mrs Chalcraft wouldn't send us a dangerous horse."

"I don't know," he replied. "Her horses were never really broken, just gentled. She could do anything with them, but others couldn't ... and he dislikes men remember."

We were to find that Stobie did not merely dislike men; he downright hated them, with a very few exceptions.

The groom led him out, a big horse all of seventeen hands. Although thin and in his rough winter coat, he resembled his photograph, even to the white star on his forehead; there was no mistake. He showed us the whites of his eyes and took a parting nip at the groom's shoulder.

"Mean as they come," said the man. "If you take my advice you won't put your wife on him."

Rather surprisingly he went willingly into the box, seemingly as glad to get away from the stable as the owner was to see him go. As we climbed back into the cab, the happy anticipation with which we had come subsided behind a cloud of gloom. We no longer took pleasure in the scenery, although the sun was setting in a splendour of fiery hues. Stobie soon changed his mind about the box and showed it by crashing his heels against the sides.

After we'd listened to this for a bit, Dug motioned to the driver to stop.

"Get out Jo," he said, "and talk to him. You're good at quietening horses."

Doing as bidden, I looked between the slats, straight at two large hooves raised in anger. I began to talk in the low pitched, dulcet tones that the boys referred to as 'mummy's soothing syrup', guaranteed to calm horses and most distraught males. The hooves went down without a crash and Stobie turned round. At least he was disposed to listen and the syrup seemed to work. The next ten miles were quiet and peaceful, then the pounding began again.

"Shall I have a few more words with him?" I asked.

"We'll get on," replied the haulier. "He just doesn't like being boxed – some don't."

I thought he sounded very cheerful considering it was his best box that Stobie was kicking the hell out of. Even so, we were all relieved when we reached the farm and Stobie was installed in the roomy loosebox at the end of the stables. He did three rapid circuits of the space then reared his head over the door with a final bellicose glance at us.

"Never mind darling," Dug consoled me, noticing my glum expression, "if he's no good I'll try and get you something else."

"Let's give him a try," I said. In spite of his behaviour I liked the big brute. He was obviously upset and on edge, no doubt due to two changes and some mismanagement at the stables.

The next morning Dug turned him out in the paddock between the buildings and the cottages and we left him for a week. It was a pleasant little pasture, plenty of grass and sheltered on the North and East side by belts of woodland. From the roadside Stobie could see much that went on about the farm. As he had been bred and reared on a farm this no doubt helped more than anything to make him feel at home. He stood for hours by the fence, head and ears erect watching the tractors at work.

He had a great liking for tractors. About the third day he was missing from the field. How he got out we never knew. We were about to organise a search when John saw him at the back of the tractor shed, standing between two machines as if they were stable companions.

I visited him every day with a tit-bit. He was fussy about them. Sugar he refused, apples would be accepted only if they were Cox's Orange Pippins. What he really preferred was a handful of linseed cake or dairy nuts.

By the end of the week he was striding to meet me and nuzzling at my jacket and taking an apple core from my mouth.

Then one morning Dug put the saddle on him and rode him up the lane.

"How did he go?" I asked.

"Perfectly. Quiet as you like." Obviously Dug was now on Stobie's brief list of acceptable males. "I cantered him along the grass verge beside John on his tractor, he's as comfortable as an armchair, you can take him out this afternoon."

After lunch Dug saddled him up for me and gave me a leg up onto Stobie's back. It felt like being on the top deck of a bus. He held his head high on his long neck so that I was looking between his ears as through the sighs of a rifle, but he had a beautiful mouth and felt as safe as a house. To my indignation Dug insisted on escorting me on a bicycle, but he was apparently satisfied with our performance and said I could take Stobie out whenever I liked.

My first ride, however, turned out to be a tussle of wills. I set off for Bignor Hill, careful not to tell Dug I intended riding so far. April was past and with it the good weather, May had come in with a cold blustery wind and Stobie, despite his Canadian dam, did not like the cold. He disliked even less the snowflakes that fell as we reached the other side of Black Jack Wood. Maybe I should have turned back, but I obstinately wished to reach the signpost. We arrived there the same time as a blizzard and whirling snowflakes blotted out the Downs.

Stobie now proved to me that he had not become entirely docile, but somehow we managed to arrive home still together.

In the years that followed we had many much pleasanter rides up to the signpost, though we sometimes experienced strange phenomena of weather. Leaving the open down in bright sunlight we would find Black Jack Wood blanketed in thick, white and chilling fog out of which, the dark bulbous shapes of trees loomed like distorted figures in a witch's forest.

Stobie, hating it, would mince along the path we could not see, shying and blowing as if a hobgoblin were hiding behind every weird and ugly tree form. Once more on the Downs we would come into welcome sunshine.

Stobie proved a companionable horse, soon knowing my mind and moods. If I wanted to stroll and admire the scenery, that was alright with him; if I felt like a good gallop, he was game for that too. He enjoyed a little conversation, although his part was restricted to the movement back-and-forth of his ears, but this was, after all, more indication of listening than I sometimes got from my family.

Even when he was obstinate or upset, I could control him as much by voice as by any other means.

However, he did have one disconcerting habit. He would not pass courting couples lying in the long grass. He stood over them, his neck craned, peering down at them in curious surprise, impervious to all my efforts to move him on. I think my embarrassment was more acute than theirs, for often they went on with their sweet pastime without so much as an upward glance.

Rather more serious was his aversion to being shod. After he had put Ron the farrier out of the stable door, despite his being built like Longfellow's village blacksmith, reared up and crashed his head against a beam, narrowly missing Dug's head with his flaying hooves as he came down and then pinned him against the manger, they decided that enough was enough. Ron took off the shoes and Stobie went unshod after that. As we always rode him on the soft, this proved quite satisfactory.

By the end of May, Stobie had settled down as if he had been with us all his life. He had filled out and with his summer coat a rich mahogany bay, dappled on the hindquarters, he began to look as handsome as he really was.

"Yer wants t' give 'im some mangolds," Fred told me. "It's just the time now, when they're soft an' withered. They'll put a sheen on is coat." I fetched some from the clamp and took a couple to Stobie. By the way his teeth went into them I could tell he knew all about mangolds.

One day as I leaned on the paddock fence just admiring him and thinking how lucky we were to have him, I witnessed an amusing scene. Stobie had his head down and as he intently cropped the new grass, his gradual steps were bringing him slowly nearer the clump of nettles where Mrs Black Stocking had her brood of chicks.

Suddenly aware of this approaching menace to her young, the little hen flew out of the weeds, all flapping wings, outstretched neck and raucous voice. Without sparing her a glance Stobie went on unconcernedly ripping at the grass. Indignant at being thus ignored, but for the moment unsure of her next tactics, Mrs Black Stocking took several short runs at this seemingly immovable mountain of horseflesh, uttering harsh squawks. Stobie appeared not to notice and went on grazing. He was now very close to the nettles.

Enraged and panic-stricken, the bantam hen took a long run and hurled herself at Stobie's head. Screaming abuse, she flew up and down his lowered neck, beating her wings against him, stabbing him with her sharp claws.

Quite undisturbed, Stobie went on grazing. This complete indifference to her onslaught deflated even Mrs Black Stocking. Utterly mortified, she collected up her brood and with as much dignity as she could muster, marshalled them to a place of safety.

In the autumn, Mrs Chalcraft came to see us. She arrived on a day of thick fog. Dug, anxious about her driving from Hampshire, set out to look for her. They met, almost head-on as she took the bend off the Chichester road at considerable speed. She was in very good spirits because the thirty-two year-old Rolls had just passed its MOT with flying colours.

Stobie was delighted to see her and I think we went up in his estimation now he knew we were acquainted with his beloved owner.

"He looks happy enough," she said. "I've never seen him look better or so gentle."

Amazingly gentle he could be; we became very fond of him and he, I think, of us, but he always remained a bit of a Jekyll and Hyde and very few outside the family could be really sure of him. I sometimes led Stobie round the field with all three of the boys on his back. Nigel could handle him alone on the lawn, although his feet would not reach the stirrups – something that would have amazed Mrs Clarcraft and the man at the stables would never have believed – but Stobie was obviously too big. We wanted something smaller that they could learn to ride, so the following spring we decided to get a pony for the boys.

Choosing just the right pony for one's children is never easy and we went through a process of trial and error. The first was Hush, a pretty grey gelding, about 12.3, which had been the adored mount and companion of an only daughter and a household pet. This should have warned us that he would be spoilt, but when I tried him in the field I encountered only a slight stubbornness, which we put down to his not having been ridden for some time.

At home he got on fairly well with Stobie, except when he threw up he heels in his senior's face and had to be put in his place with a sharp bite in the rump.

Although still stubborn, when I took him out alone he went well in the company of Stobie and was well behaved when I led him round the lawn or paddock with one of the boys on his back.

Evelyn had left us earlier in the year to marry Ron the farrier and now that boys were getting older I decided to look for a girl who would help with the horses and the boys riding lessons as well as the domestic chores. Marie was slim and attractive with brown curls, wide green eyes and a dusting of freckles across her small nose. She was quiet and capable with the boys and the horses; even Stobie approved of her.

Hush meanwhile, with the good grazing, was getting increasingly lively and decidedly fat. He put on weight until he looked like a Thelwell drawing. Even keeping him in during the day made very little difference. His rotundity did not make him any easier to sit on and then he developed a bad habit of bucking.

I rather enjoy a horse bucking so long as I can feel him doing it. It is when I feel nothing until I hit the ground that it ceases to be fun. Stobie, when he was feeling good, often started off at a canter with a flying buck and if another horse galloped up behind him he would throw his heels so high that my head touched his tail.

Hush's bucking, too, was fairly predictable. Marie and I had no trouble in sitting them, but for the boys just starting to ride he clearly would not do. We sold him to the riding establishment of a girl's school near Chichester. With plenty of regular work he kept slim and docile and remained a great favourite with the school until a ripe old age.

We got Rocky through an advertisement in the local paper. 'Reliable children's pony' it read; 'Cheap to good home'.

He was eleven years old, black, 12 hands, with a long bushy mane and tail and trim slender legs. We liked the look of him and this time we made no mistake. There was one surprise. Although reliable and quiet with the boys, whenever Marie or I rode him, he could be keen and very fast.

He and Stobie were friends from the start and stood nose to tail in the field, flicking away the flies like two old carthorses.

We congratulated ourselves that we had made a good purchase. Nigel, by this time was off the leading rein but showed a disappointing lack of enthusiasm for riding. There were so many other things to do. Only when rain teemed down and it was too wet for any other outdoor pursuit would he ride. There are things I would rather do than ride in the rain, but that summer I spent many long, wet hours in the saddle accompanying Nigel.

I had hoped that I might exercise horses and dogs together but although Beauty, now well grown, was keen to follow behind the horses. Jane would only do so for a short distance and then sit down in the middle of the road, looking wistful but adamant. She would watch us out of sight and then turn back. At ten years old she was a bit past horse exercise.

One day Nigel came home from school with a small corgi at his heels.

"He was wandering about in the village," he said, by way of explanation. "When I spoke to him he followed me."

Clearly the dog was lost and hungry. I fed him and Jane hospitably allowed him to share her basket.

The next morning I telephoned the police and the RSPCA. The dog wore a collar with a name Rusty, but no address. After breakfast he followed Dug into the yard, jumped into the Land Rover and remained there all the morning as if this was what he was used to. When not in the Land Rover he attached himself to me, following close at heel on walks, around the house and garden and when I fed the hens.

Days passed, I repeated the telephone calls and put an advertisement in the Lost and Found column of the local paper. No one seemed to have lost a corgi.

"Do you think we'll be able to keep him?" asked Nigel.

I shook my head. To keep a dog with Jane and Beauty would not be practical.

"Then if no-one claims him, what shall we do with him?" he inquired, anxiously.

The question had been nagging at the back of my own mind. Rusty had slipped quietly but firmly into the routine of our lives and into our hearts. He might have been with us all his life. We knew that the longer he stayed the more difficult it would be to part with him. Already the idea of turning him over to the police or even the excellent dogs home only a few miles away was hard.

Then one evening a young couple came to the door.

"We understand from the R.S.P.C.A. that you have a stray corgi," they said. I asked them in and went to fetch Rusty, who was romping in the walled garden with Beauty. His rapturous greeting of the pair left no possible doubt that they were his people.

"We live in Littlehampton, but Rusty's home is in Yorkshire," the young woman told us. "He belongs to my father who was on holiday with us. Rusty got out and was lost. I suppose he was trying to find father's farm." So he had walked the eight miles from Littlehampton when Nigel found him, quite a way on his little short legs.

"Father had to go home without him. He was broken-hearted."

"We thought he must belong to a farm," I said. "He was so at home here. Does he ride in your fathers Land Rover?"

"All the time," they replied.

They thanked us for our care of Rusty and rose to go.

"You didn't see our advert in paper," said the young man. "We offered a £5 reward, we'd like you to have it."

"It was Nigel who found him," I said. "I think it should be his."

Nigel was pleased to receive the £5. The following Saturday he spent it on a comfortable new bed for Jane.

We had thought that one advantage in having a second horse was that Dug and I could ride together, but somehow the summer had passed without our finding time to do so. Then one Saturday afternoon at the end of September we saddled Stobie and Rocky and headed them towards the Downs. In the woods, the oaks were turning yellow and the beeches beginning to take on their autumn hues. Across the Downs stacks of straw bales and wide stretches of golden stubble signified the end of one farming year, while long brown stripes carved by the plough heralded the beginning of another. Scarlet hawthorn berries beaded the hedgerows topped with feathery trails of old man's beard. Pethwine, John and old Fred called it, the Sussex name, and told me how when they were boys they used to smoke its dried hollow stem like a cigarette and thatchers used the long wiry stems to tie on the yelms – a Sussex term for straw sorted and bundled for thatching.

We paused to watch a kestrel hovering, its great brown eyes seeking out a mouse in the stubble. Then it plummeted down upon its prey. It was when we moved on that Rocky, who up to then had been walking out with his usual zest, began to behave oddly. At the slightest pressure of my heel on his right side he would whip round in that direction. I dismounted and felt his side. There was a definite tenderness.

"We'd better go back," said Dug.

I loosened the girth and led Rocky, who soon showed signs of being in intense pain. We put him in the loose box, rugged him up and sent for the vet. It was an intestinal virus, he said, and gave him an injection. For several days we had a very sick pony on our hands but with a course of injections and careful nursing he began to show some improvement. It was a long convalescence. At first he spent the nights in the box and on the lawn by day. Presently the vet advised a little exercise and we took him out each day on a halter when we walked the dogs. By the end of October he was almost fit and we looked forward to soon being able to ride him again. One Saturday we took the boys to visit friends in the New Forest, leaving Marie in charge of the animals. We thought on our return that she looked rather pale and upset but we were not prepared for the reason. Rocky was dead.

Marie had taken him out in the afternoon as usual and he had seemed all right. She had left him to have tea and feed the dogs. When she next went out to the stable he had died.

We had not the heart to look for another pony and the boys did not ask for one. They had ridden, more because riding had been organised for them than from a real desire. They had other things to do.

8. Other Horses and Au Pairs

Marie left us at Christmas. She wanted more work with horses, she said, but I suspected that a secondary reason was the winter in our cold house. During a spell of frosty weather in November she had performed her round of the bedrooms wielding a mop and wearing her hacking jacket, gloves and scarf. She found a job at a stables within easy distance of her own cosy, centrally-heated home. We were sorry to lose her and as temperatures dropped well below zero in January, I thought of her with faint envy.

The schoolgirl daughter of our herdsman came in daily during the spring, but when Easter came round I wondered how I could cope with cooking and cleaning, the garden and still occasionally exercise the dogs and Stobie.

"Why not get an au pair?" someone said.

While I was toying with this idea a friend phoned me. She had a German girl with her for a month, an excellent worker but of uncertain temperament. My friend was finding her difficult to deal with and her visa had another two months to run. Would I like to give her a trial?

Now I have never been daunted by a little temperament. I invited my friend to tea, bringing her au pair, so that we could see how we liked each other.

The result was that Hansi came to us early in April. She was a strong, dark-haired, handsome girl of seventeen. Her forceful character was revealed in the direct gaze of her clear grey eyes, her firm mouth and purposeful walk, which was accentuated by the short leather boots she wore.

She was highly intelligent, quickly mastering English and had a sense of humour, which although at times could be misconstrued, was also contagious and irresistible.

As for work, she was the proverbial 'new broom' and set about cleaning the house with a zeal and vigour that frankly terrified me and reduced the boys to a state of dumb amazement, too shaken, for a while at least, to complain about her clearing up forays, in which all their treasures were ruthlessly tidied away, sometimes never to be found again.

By the end of a week, a transformation had taken place from attic to wash house. In place of a jumble of muddy boots and shoes in the latter there were neat regimental rows of clean, highly polished footwear even my old gardening shoes, caked with the mud of many seasons, looked fit to walk out in.

The silver and cutlery was not only shining brightly but also arrayed in faultless head to tail formation. Every bit of the floor was swept and scrubbed or polished, even to the remote places where I usually swept the dirt.

Hansi clearly despised my efforts to keep the house clean. Fortunately perhaps, her muttered opinions on the subject were made in German, although 'swine houzen' needed no interpretation. But if I could turn a blind eye on occasion I could also turn a deaf ear and her remarks did not impair our relationship.

The main problem seemed to be her preference for organising her work herself and in a somewhat unorthodox manner. Some mornings an hour would pass while she picked and arranged flowers, a job I liked to do myself, and put them in every room, even the smallest. Sometimes she would be missing and I would find her grooming Stobie, or riding him, but when Dug came in at the end of the day to sit down and relax in his favourite chair, he would find it out on the lawn with much of the other furniture and Hansi busily vacuuming the carpet.

Here we drew the line. Stobie had dealt with her high-handedness in his usual way of sweeping her from the saddle under a low bough. For several days afterwards she walked with a pronounced limp upon which neither of us remarked, but I suggested that if she would do the housework at a time convenient to the family, I would accompany her on foot if she wished to ride. I knew that with me along, Stobie would behave himself. This worked very well. The rooms were turned out in the mornings and riding instruction took place in the afternoons. Hansi was a natural horsewoman and learned quickly and I found that keeping up with Stobie's long stride and trot across the Downs was excellent for the figure, my own and the dogs.

Now that Hansi relieved me of more domestic chores than I knew existed, I had more time to spend on Jane and Beauty, especially Beauty. For some time, friends and acquaintances in the dog-breeding world had urged me to show her. If one possesses a good specimen of show strain it was nothing less than a duty, they inferred, to exhibit it. So I entered Beauty for the Cheltenham Show to be held the following July.

Meanwhile there was a lot of work to be done on her and my more knowledgeable friends were very ready to help and advise, how to trim her ears, feet, legs and tail; the finer points of grooming; how to 'show' her. I spent some time every day walking her on the road, trotting her up and down the lawn and teaching her to stand correctly like a debutante before her presentation.

When July came I noted with satisfaction that she was looking her best with shining black nose, bright eye and a glistening full coat.

She was a different type from old Jane, who had witnessed all this fuss over Beauty with a seemingly bored indifference. Her coat was smoother and finer and much lighter, the colour of crushed maize fading to a pale cream at her under parts and she was of a less sturdy build. By this time Dug was busy with the harvest, so two friends, experienced in breeding and showing fox terriers offered to take me to Cheltenham. It was a beautiful day and would have been a pleasant drive but for the object of the journey. Beauty was prone to carsickness but never so much as on this occasion and we had every need of the sacks and pile of newspapers we had brought.

My friends showed the utmost patience at our frequent stops. The day grew hot and it was with relief that after several hours of travelling we arrived at a lovely area of parkland just outside Cheltenham where the show was held. We parked the car and led out Beauty, who looked extremely sorry for herself.

After the vet had passed her, I was shown her section of the bench in one of the marquees. I gave her water and groomed her and then left her, still looking peeved and dejected in the companionship of a sturdy Labrador on one side and a disdainful English Setter on the other, I went to look round the show ground while awaiting her class.

There was every breed of dog from minute Chihuahuas to Great Danes and stately black and tan Gordon Setters. Most of their owners looked far more confident and expert than I felt.

When the entries for the Golden Retriever Maiden Bitch class began to gather, I went to fetch Beauty, hoping that she would have cheered up a little. She had not and when her turn came to parade in front of the judge, she slunk past him like a sheep to the slaughter.

I knew that however show-worthy she might be, the judge would be looking for a little 'joie de vivre' and I would have lost marks there. I should also have known that you do not stand admiring the scenery allowing your dog to doze at your feet while he is looking at a rival entry. At any moment he will pass an appraising eye round the whole class and you must be on your toes, showing your dog for all you're worth when this happens.

Concerning Beauty's and my performance, the judge had only one comment to make on his card: 'Handler and dog totally inexperienced'.

Later I met Beauty's breeder.

"It was a pity you did not realize you had to be showing all the time," she said.

I could not argue with that.

"And do get her a show lead," she added.

I had noticed in the ring that an ordinary collar and lead, however well cleaned and polished for the occasion was not what the well-turned out dog was wearing.

My nose, following the rich aroma of new leather led me to the stand of leather goods. There I brought an expensive show lead in soft whale hide. It was rather like buying a dress the morning after the ball. It was a pity I would probably never use it, for something told me that my debut into the dog showing world was also my swansong.

The whole thing had been too much for Beauty. She slept soundly all the way home.

My friends dropped me at my front door. No, they would not come in. They looked as tired as Beauty and I felt that I had sadly disappointed them with my incompetence.

Inside the house I was met by an aroma like that of a Soho restaurant and found Hansi in the kitchen. She had concocted a dish that was more to her liking than our English food, which, she declared, "tastes for nothing" (she still had trouble with her prepositions). It consisted chiefly of a species of fungi which I had always thought to be inedible, flavoured with herbs and garlic. Nigel and Peter, always conservative in their eating habits, had refused to partake of this delicacy, but Nick was tucking in with keen enjoyment. Although I was aware that more types of fungi are edible than is generally known, I still felt apprehensive.

"Are you sure they are not poisonous, Hansi?" I asked.

"Ah, but they are not ze poison that kills you," came the reply. Not too reassuring, I thought.

In a cardboard carton on the table was a very unhappy looking hedgehog and the remains of a saucer of bread and milk.

"I found him in the garden, he is sick," Hansi explained.

Since her arrival she had outdone Nicholas in bringing in sick and wounded birds and animals for treatment. I suggested that the hedgehog was merely suffering from indigestion after a surfeit of bread and milk and that with exercise and a return to its natural diet it would recover.

At this Hansi swept up the carton and carried it off to her bedroom, her feet in their short boots resounding on every stair. In face of this indignation I did not like to remark that a bedroom was not the most suitable place for a hedgehog and its many attendant fleas.

Apart from a few slight differences of this kind, Hansi seemed happy with us. At the end of the summer she applied for an extension of her visa and was still with us at Christmas.

We had tended to collect animals when in Hampshire. In the autumn we visited a friend, Michael, who was managing an estate near Winchester and his wife Ruth.

As we drove slowly up the gravel drive to their house, a pony cantered up to the fence and regarded us with interest. We stopped to look at him. He was a handsome, dapple iron-grey gelding, about 14.2, with a neat head and wide intelligent eyes. I could see that Dug was as taken with him as I was. We were not exactly looking for another horse, but we'd had it in the back of our minds for some time. Stobie was twenty-one years old and though in good condition generally and often surprisingly lively, he was sometimes lame for two or three weeks at a time with a permanently swollen fetlock joint. He had always to be ridden with extreme care. Besides it would be nice to ride with Dug again and Hansi, whom I still accompanied on foot.

"Nice pony that," Dug remarked to Michael and Ruth after we had greeted them, nodding toward the grey which, had continued to follow and now stood watching us from the other side of the fence.

"He's half Arab, and for sale actually," Mike replied. We learned that he belonged to the owner of the estate and had been hunted by the young son who had outgrown him. "I'll make a few enquiries for you if you like".

The result of Mike's enquiries was that Sleek came to us a few weeks later on a month's trial. Now the object of having a horse on trial is that you return him if not satisfactory, at least, any ordinary sensible person does. But my natural optimism seems to run away with me when buying a horse.

I had a feeling from the time that Sleek arrived, beautifully boxed and wearing a smart, monogrammed blanket that we were not meant for each other. My first ride should have confirmed this.

"Now take him quietly," Dug advised me as we set off.

The last thing Sleek wanted was to be taken quietly. He danced and chafed at the bit, but I managed to keep him at a sedate walk or trot, except for one short gentle canter, until we reached the last strip of wood on the way home. Suddenly he threw up his head, caught at the bit, tucked his hind quarters under him and went flying along the narrow, uneven path, past jutting bushes and low overhanging boughs like a Derby winner up the home straight.

I did not quite manage to pull him up before we reached the road and almost collided with Shep, who was pushing his bicycle.

"Bit lively ain't 'ee?" he remarked with nice understatement. "You wants good bit on 'im. Snaffle's no good."

I was inclined to agree, but when I said so to Dug he scoffed.

"Trouble is you women don't know how to manage horses. No horse needs anything more than a snaffle just hacking."

I had known other horsemen who subscribed to this theory, which sounds very good and sensible if you are not sitting on you hack when it is bolting at about thirty miles an hour, its nose pointing skywards and the bit between its teeth. Sleek's previous owner had ridden him hunting in a Pelham and standing martingale and I would have been happier if he had something to stop him getting his head in the air.

My next few rides were uneventful and except that after Stobie's massive neck and shoulders I felt, on Sleek, that I had very little in front of me, quite enjoyable. Meanwhile he was endearing himself to other members of the family with winning ways. A whistle from the front door would bring him galloping to the field gate. There was never any need to go after him with a halter, he would follow us into the stable. He sometimes walked through the hall, if the front door was open and into the kitchen, nuzzling for a tit-bit. At the end of the month we decided to keep him. He had only bolted once more in that time, after a quick whip round at the sight of Shep's bucket in a hedge. I persuaded myself that he was going to be all right.

"He's just the mount for you," Dug affirmed. "Suits you a treat, better than that big old thing."

But I knew I would never attain the empathy with him that I had with Stobie. I couldn't match his enthusiasm and felt rather sad that I never rode Stobie now, all my time being taken with exercising the energetic Sleek. He stood by the fence watching us go off, looking puzzled and doleful.

Just before Christmas Stobie developed a badly infected foot, the one with the swollen fetlock. I did my best for him with bran poultices and pink ointment, tearing myself away as frequently as possible from baking mince pies, wrapping gifts and all the other festive preparations, to minister to him in the stable. Finally the pus dispersed and the inflammation subsided, but he had lost condition badly and the joint was enormous. I nursed him carefully through the bleak days of January but he did not improve, although when I let him out on the lawn for a bite of grass and exercise he would surprise me by leaping off the bank and throwing up his heels from side to side as he cantered round the lawn. But this was only the challenge of his old spirit. We all knew that he had come to the end of the road. One evening I talked on the telephone with Mrs Chalcraft. She agreed.

"I thought when I saw that fetlock joint," she said, "that it wouldn't be long."

The next morning I went out to catch Stobie in the field. I had always been the only one he would allow to put a halter on him but he usually liked to play me up a little first, not running away but extending his head away and slowly turning so that he remained just out of reach. But this day, for the first time, he came to me and put his head down. There was a man waiting in the yard and a lorry. I handed the halter rope to Dug and before I reached the house it was all over.

Sleek knew. All that morning he cantered up and down the field or stood whinnying at the gate until I could stand it no longer and shut him in the stable. Stobie's going unsettled

Sleek. He was restless in the field, chasing like a mad thing every car that drove up the hill and every bird that flew overhead. He even chased aeroplanes.

He hated being in and was increasingly unpredictable at exercise. There were lambs again on the Downs and at the sight of them gambolling by the field fence he would whip round and gallop in the opposite direction, a pheasant flying up under his feet had the same effect. Furthermore, he developed a habit of stopping dead in his tracks, his head, which, he had held almost back to my chest now thrust between his knees, so that on two occasions 1 was wrenched from the saddle. After a long period of dry frosty weather the ground was hard, but I survived with only a headache and a badly bruised shoulder.

I remembered Mrs Chalcraft telling us of an Arab horse she once had. When mounted by anyone it disliked it would gallop to the nearest hedge or gate and stop dead, depositing its rider on the other side. Perhaps this behaviour is typical of the breed.

About this time, Hansi, who had now been with us more than a year, began to grow homesick and it was arranged that she should return to Germany in June. During the last week of her stay I was painting out the larder and pantry, hoping to get the job done before she went. As I slapped away with my brush I could hear Sleek restlessly circling his box. I had not ridden him for several days and the abundance of grass meant that he must be kept in for most of the daytime.

Presently came the clatter of his heels against the door, more rapid circling then the sound again. He was demanding exercise in no uncertain way. Reluctantly, I put the lid on the paint pot, cleaned my brush and went up to change.

Sleek was on his toes as we went up the lane, peeping and shying at the most commonplace objects. When we reached the opening into the woods, just past the cottages a tractor emerged. Sleek span round and bolted towards home. I pulled him up before we reached the end of the paddock fence, turned him and pushed him on until we again reached the opening. The tractor driver, who had withdrawn with his machine at the sight of Sleek, chose this moment to pull out again.

Although prepared, I was unable to prevent another lightning whip round and this time there was no stopping him. We careered past the cottages and the paddock. As we passed the yard gateway my hard hat, which, I had recently taken to wearing, bowled away, but I felt I was beginning to wear him down. Another ten yards or so and I would be able to pull him up. Just then two men ran out from the other end of the yard to stop him and stop he did, dead in his tracks, with his head between his knees.

Ten minutes later Dug was driving me to hospital. In the passenger's mirror I took a tentative glance at my face, which had taken the brunt of the fall. Blood spouted from a deep gash over my right eye, while from nose to ear and brow to chin very little skin had escaped cuts and deep abrasions. I looked as if I had been several times through a car windscreen.

"Are you alright?" Dug asked many times as we covered the miles into Chichester.

"Mmm," was all I answered. I could not tell him that the pain over my eye and in my jaw was excruciating and that my vision was gradually going. In casualty I was wheeled away and parked in a siding to wait for what seemed an interminable time for attention, while poor Dug paced the waiting room like an expectant father.

At last a young doctor appeared at my side and things began to happen. First a local anaesthetic while five stitches were inserted in the main cut and the rest of my face cleaned up. The eye was now completely closed but the sight in the other had cleared, then an x-ray of my head and jaw.

"Mrs Duggan what haf you for bone?" exclaimed Hansi later, when she heard that nothing was broken.

Lastly two tetanus injections at a half hour interval and a cup of tea and I was ready for home.

As we walked up the drive Sleek was grazing by the field gate. He raised his head and looked surprised to see me. The next morning Dug took him out and put him through his paces.

"Good Lord, Jo," he said afterwards, "I had no idea what you were riding. The devil nearly had me off twice!"

I said nothing, wondering how we were going to get rid of him, for go he must; he and I were through.

It took several days for the pain to subside, meanwhile my face assumed the proportions of a pumpkin. Because there was not enough room on the one side for all the swelling it spread across to the other, partially closing that eye. By holding my head at a certain angle I had a beam of vision like looking through a keyhole. Before the right eye was a constant black and white rain.

"Hmm," said my doctor, when I told him. "I hope you are going to see with that eye when it opens."

I hoped so too.

For four days I lay in bed, luxuriating in the life of an invalid, while Dug, Hansi and the boys waited on me and my friends called with flowers, condolences and even grapes.

Then it was time for Hansi to leave and I was back at my place at the kitchen stove. There were certain dishes I often said I could cook with my eyes closed; now was my chance to prove it.

But by the end of the week both eyes were open and I could see perfectly with each. My doctor was concerned, however, that I had no domestic help.

"Can't you get another au pair?" he asked.

So we contacted an association, which more or less guaranteed an au pair by return of post. Sure enough within a few days a Swedish girl was on her way to us.

Zoe looked very pleasant in the tiny photograph accompanying the form we received. That evening Dug and I waited on the platform of our nearest railway station to greet her. The train, when it arrived, was half empty, a few businessmen got out, two women and a tall blonde of about twenty. We looked at her admiringly and then beyond her for Zoe. There was no-one else getting off the train. Then the tall blonde came to us with a diffident smile. There was no doubt now, this glamorous Britt Ekland cum Anita Ekburg of a creature was my au pair.

As we shook hands and I looked at her flawless complexion and wide, long-lashed blue eyes, I was conscious of my own still badly marked face. Although the swelling had now gone and it was healing well, it was far from a pretty sight.

As Zoe's command of English was limited, so was our conversation on the way to the farm, but she was able to convey her appreciation of the scenery as we drove under the verdant tracery of the beech leaves.

"How beautiful," she exclaimed, "so green. In Sweden it is not so green."

She was charming as well as decorative but one thing bothered me. How was I to set this exquisite creature to work in the house? We learned that her father was with the Swedish Consul in London. That evening he telephoned to assure himself of his daughter's safe arrival. Then he told me that as there had been cases of smallpox in England (we had heard of none) he had thought it advisable for Zoe to be vaccinated. This had been done a few days before she left Sweden, so she might not be feeling very well. Would I please give her only very light work for the first week.

Now I really intended to comply with this request, but unfortunately Zoe's idea of light work and mine were poles apart. Of preparing vegetables, she had no idea. Vacuuming the carpets she found most strenuous.

"Its so hard, so hard," she complained. "In Sweden we do not work so hard."

I could have said that there were smaller, more modern houses in England where we do not work so hard.

"What do you do at home?" I asked.

"I study. That is hard work."

"And when you are not studying?"

She opened her eyes wide that I should expect further exertion.

"I sleep," she replied.

Each evening as she talked to her father on the telephone her voice became more and more distressed. One evening he asked to speak to me. His tone was stern and reproachful; obviously I was not looking after his daughter. She was unwell, he said and he would need to be seen by a doctor. Certainly that evening she had complained of pain in the vaccinated leg and looked flushed and tearful.

"Would you like me to take your temperature?" Dug asked her.

At this her eyes opened wide.

"No! Oh, no," she exclaimed, with what seemed unnecessary vehemence. We did not press the matter.

The next morning I was taking no chances. I kept Zoe in bed and took up her breakfast. Then I called the doctor.

When he came about midday, Zoe extended a long, shapely, suntanned leg to show him the single distasteful blemish.

"How long will it be so?" she asked, her huge eyes brimming with tears.

I thought I detected a glimmer of amusement behind the doctor's grave manner.

"Only a few more days," he answered gently. "It is a perfectly normal reaction, but I had better take your temperature."

Taking out his thermometer he held it to her lips.

"In ze mouff?" she exclaimed in astonishment.

I looked at the doctor and again perceived that almost imperceptible twinkle. Only then did I understand Zoe's acute embarrassment when Dug had offered to take her temperature the previous evening.

As I saw him out the doctor raised a quizzical eyebrow.

"This is your help?" he asked.

Zoe came down later and as it was a day of warm, cloudless sunshine Dug suggested that she should sunbathe in the garden.

"Ya Ya," she replied, with the first smile we had seen for days and went upstairs again to change. She came down looking as if she had stepped off the cover of a glossy Swedish magazine, in brief blue shorts and bikini top, a matching scarf tying back her blonde hair.

I found her a rug and she lay on the lawn while the tractors and Land Rovers passed the gate with what I thought was unnecessary frequency.

That evening Zoe and her father arranged that she should join him in London the following day, so just a week from the time we met her we were seeing our au pair off at the station.

"Goodbye," she said, "and I thank you," then with her most disarming smile, "I am so sorry."

We knew that she was genuinely sorry that she had not been more help to me. She was a very sweet girl. The fault, if there was one, lay with whoever sent her to a farmhouse.

A few days later I heard from the au pair agency. They were sorry that things had not worked out with Zoe, but they now had a Dutch girl on their books. The Dutch, they said were very hard working. I thanked them, but declined.

It was not so difficult after all to find a home for Sleek. One day a vivacious teenage girl came to the door, she had heard that we had a horse for sale.

We took her to look at Sleek in the paddock. It was a case of love at first sight and this time it was mutual. Here was a rider who was as young and wild and energetic as himself. We warned her of his past behaviour and insisted that she had him on trial, but it was soon apparent that they were made for each other.

At the end of the month she bought him complete with tack and we saw them frequently galloping across the Downs early in the morning and in the evening. At last he had all the exercise and companionship he needed.

Sleek's going was virtually the end of horses for us, but there was a postscript. Shep wanted a pony and cart to carry hurdles and fleeces and food for the sheep to and from the Downs. Very soon Dug bought a smart but solid little trap from the local gypsies. It now remained to find the right pony.

The wife of a local farmer had a mare for sale, quiet to ride and drive and offered her to Dug on trial. She was a cobby type about 15.2, an attractive flea-bitten grey. She came from Devon and her name, appropriately, was Widdicombe.

The day after her arrival Dug attempted to try her out in the trap. The farmer's wife told him that Widdicombe went well in harness, but no one it seemed, had told Widdicombe. The only way she would go was straight up on her hind legs immediately she got between the shafts. Dug decided to try her again the next morning with some help from the men.

The next morning he was so late in to breakfast that having got the boys off to school, I went to see what was happening.

I found John and old Fred by the cart shed.

"Lookin' fer t' guv'ner?" asked Fred. "He's gone flyin' up t' road behind t' horse."

I sincerely hoped there was a cart under him.

When at last he came in I heard the whole story.

With the help of John and Fred he had backed the mare into the shafts and fastened the traces. They had held her while he climbed up and took the reins. When they let her go she had reared up as before, but after a crack from the whip she broke into a gallop, the cart rounding corner of the yard on one wheel and pounded along the road to North Wood.

Dug had stopped her by turning her into a ploughed field but one rein had got caught under the shaft. Widdicombe backed rapidly, narrowly missing trees to the right and left. When there was room ahead, Dug let the rein go and it fell free. When John and Fred came to look for him he was driving sedately back to the farm.

"She no good for Shep, that's certain," he said. "Now she'll be here for a month, eating her head off and doing nothing."

"I could ride her," I said.

Dug looked doubtful. Widdicombe had not so far acted up to her reputation of quietness and I was still bearing the marks of my last ride.

"Just because she didn't like the cart doesn't mean she isn't sensible to ride," I persuaded him.

"All right, try her. Stobie's tack ought to fit her."

She seemed a bit skittish when I mounted her the next afternoon, but quickly settled down. The set of her little black-fringed ears and the way she had of peeping reminded me of Sleek but she was a much steadier mount. We rode out

116

through the woods and cantered back along the Downs. We were completely happy with each other and by the time we were back in the stable she had fully restored my confidence.

I rode her often during the following weeks and was sorry to see her go. The gypsies turned up one evening with a small but strong chestnut pony, which was just right for Shep. His name was Tomboy, almost red in colour with light mane and tail he was a pretty picture in the bright green cart and with Shep sitting up in front sucking his pipe they created a rural scene that was a pleasant change from tractors.

9. *The Domestic Life*

From this time a change came over our lives, mainly mine. We decided that horses had become more of a problem than a pleasure. Now that the boys were older, I could dispense with regular domestic help, especially as Nigel was going away to school in September. From now on I would devote all my thought and time to my family and a life of complete domesticity, with the possible exception of the dogs.

Writing would have to go too. I had enjoyed a long run of reasonably successful freelance work and somewhere I had read that all writers should stop sometime and absorb in order to write again. I was now due for a period of absorbing. As a gesture towards this end, I consigned to the garden bonfire all my unpublished stories and articles, something that I afterwards regretted, for added to the conflagration was a 75,000-word manuscript written during the war years about my life in the Land Army, most of it penned painstakingly by the light of an oil lamp in my cottage billet. I later met a friend whom I had not seen since those days.

"Do you remember that book you wrote?" she asked. "It started, 'I like pigs'." I remembered the book, but had forgotten the beginning and wondered how I went on from there. Now I would never know.

Earlier in the year we had made several excursions round Sussex and into Surrey, looking for a suitable school for Nigel.

The one that impressed us both was not far away, just this side of Worthing. Situated in a beautiful and well-kept house and grounds once belonging to a member of the Bowes-Lyon family, we were pleased not only with the surroundings but with the curriculum and general standard of the school.

During the last weeks of August, shopping for uniform, sewing on nametapes, packing and general preparation for the first term was added to the usual seasonal business of preserving the last fruits of summer.

Nigel occupied himself with packing and repacking his tuck box with all his favourite sweets and goodies, something that took his mind off the less happy side of his impending departure. At last Sunday evening arrived, and accompanied by Nick and Peter, awed into an unusual silence by the occasion, we drove him to Goring.

As we stood in the hall with other 'new' parents with forced bright smiles and apprehensive small boys with brave faces, I wondered whether the Americans, who regard us as heartless for packing our children off to school, could possibly be right.

My thoughts were broken by a large formidable person who could only be matron bearing down upon us. She pointed a finger like a forbidding angel.

"No tuck boxes allowed," she boomed.

We all shrank several inches. I dared not look at Nigel but motioned to Nick and Peter to carry out the offending article.

As her appearance would suggest, Matron was a strict disciplinarian, meting out penalties for such lapses as an improperly stripped bed or a dropped towel. As we got to know her better, we found under the stern exterior a warm and generous nature, a quick wit and a deep understanding of the problems of both boys and parents, although not until she retired some years later did we realise, perhaps, just how much she contributed to the character of the school. It was never quite the same afterwards.

The day after Nigel went, we couldn't find Beauty. Every morning she and Jane accompanied Dug when he went to the dairy for the milk, chased the cats, helped themselves to a piece of linseed cake, greeted the men and returned to their baskets. This morning Beauty did not come back. All morning I whistled and called at the doors but she did not come.

Then Dug saw a golden blob in the matching stubble at the far end of a field. Beauty came when called, but went straight back. In the evening she came in for her meal but returned to the field. When it grew dark she still would not come, so the boys and I took a torch and went to fetch her. She was sitting where she had sat all day; beside her was Nigel's belt.

Why, we wondered, had she not brought it in, instead of mounting guard over it. Did she think Nigel would come back for it and was waiting for him? It was not the first time I was caused to wonder what goes on in a dog's mind.

Judging from his letters, Nigel had settled down well to boarding school life. Nick and Peter revelled in having only one female, me, to contend with, and Dug, I knew appreciated being on our own. I too found a satisfaction and freedom in doing all my own work and having the house to myself during the day.

As any housewife who has lived in a rambling farmhouse will know, it is not the work that takes so much time, it is the distance one walks in doing it. There were the forty paces to and from the larder and the same number out to the garden and the treks to the top of the house.

Interruptions took up a good deal of the day. The telephone is no respecter of a well-planned morning. There was always an urgent message when there was no one around to give it to. When a lorry driver wanted to know where to put his large load, the farm was always as deserted as a ghost town.

There were constant cups of tea and coffee for the men and the resulting endless batches of dirty beakers on the draining board, then afternoon tea to be got ready for men in the field.

"Aren't you lucky," said another farmer's wife to me one day. "Dug takes the tea out to the field. I have to get in the car and take it myself."

I suppose everything is relative.

There were snacks to get and on occasion meals for casual labour, for reps and any odd bods who needed them. One day Dug rushed in and asked me to take three cups of tea and three pieces of cake to three men in a lorry. It was not until I got out into the yard with the tray that I discovered what he had failed to notice. The one in the middle was a large Labrador dog. He enjoyed the tea and cake all the same.

Then there was the first aid. Men with fingers caught in a plough or other machinery, various lacerations; stray children with grazed knees... I learned always to first sit the patient down and treat for shock, after a large, tough-looking lorry driver had sprawled at my feet in a dead faint while I was bandaging his cut finger. Nigel came home at Christmas, taller, more confident and happy.

Maybe we are right after all and not the Americans.

Snow fell during the holidays. Our boys and the rest of the gang were out in force tobogganing all day on the sloping field by the house. By mid morning they were thronging the kitchen until it looked like Woolworths on a Saturday, sipping hot drinks, while I searched through cupboards and drawers for a fresh supply of dry gloves, socks, scarves and spread the wet ones round the fire to dry in time for the afternoon session.

By the end of the holidays I felt a bit jaded and the house looked as if it had been hit by a cyclone, so I was very pleased when Dug said the wife of Paddy, his new cowman, wanted occasional work in the house and would come in and help me straighten up.

I had made Paddy's acquaintance when he had come to the back door to ask me to telephone the vet or AI man. He was small and wiry with pale blue eyes behind steel-rimmed glasses and a soft County Cork accent. Extremely cheerful, he always stopped to eulogise about the village and its inhabitants.

"'Tis a beautiful place to be sure and I've niver met sich lovely people. Ye feel ye've known them all ye life, not like some places where no one takes any notice of ye at all."

I said I was glad he was so happy here and hoped that his wife was too. Mrs Paddy came for two hours three times a week. She was small and wiry with pale blue eyes behind steel-rimmed glasses and a soft County Cork accent. She and Paddy could have been brother and sister. She worked well if I kept out if her way but was a compulsive talker and while she talked neither of us could get anything done. So having escaped from her first burst of chatter on arrival and a second while she drunk her elevenses, I discreetly avoided her.

Left alone she quickly brought about a transformation in the house. It was good to be able to see through the windows again, to have shining furniture fragrant with polish and clean brass and silver.

Then, after several weeks had passed Mrs Paddy arrived without her cheerful smile and bright chatter. She seemed worried and preoccupied.

"Something wrong Mrs Paddy?" I asked.

"It's Paddy," she said. "He's so depressed and off his food, not himself at all."

"What's the trouble, is it the job?" I questioned.

"Tis partly the job and partly the village folk, he just dosen't seem happy with anyone."

This seemed a strange turn of events. What had happened to all those lovely people?

"He does get like this at times," she added. "It'll no doubt pass off."

On her next appointed day Mrs Paddy failed to appear. I was about to dish up the lunch when the telephone rang. The voice at the other end of the line had an unmistakable Irish accent.

"This is Paddy," it said. "I'm going to commit suicide."

"Now Paddy, you don't mean that," I replied.

"I do that. I'm goin' to take me life."

I felt certain, somehow, that he meant it. Pressed suddenly into the role of a Samaritan I groped for words.

"But why Paddy? I thought you were so happy here."

"Ah, tis the same everywhere, folks don't care about ye."

Keep him talking, I thought, don't let him ring off. I could hear a saucepan boiling over and smelled the acrid smell of cabbage burning.

Paddy talked on while I emitted sympathetic noises and the occasional remark. But perhaps sympathy was wrong, perhaps something more bracing was the right line. I wished that Dug would come in, I had more faith in his ability to handle tricky situations.

"Paddy don't do anything desperate, promise me," I said at last. "Come over this afternoon and talk over your problem with us. We'll do everything we can to help."

"Alright, I'll promise ye, I will that, blessed angel that ye are."

The blessed angel put down the receiver and fled to the kitchen on winged feet to salvage what remained of dinner.

Dug was a bit sceptical about Paddy's dire intention, but said he would talk to him and try to get him into a better frame of mind, but Paddy did not come that day or the next, nor was there any sign of Mrs Paddy.

Just before lunch the following day I answered a pounding on the front door to find a man, white-faced and trembling, his eyes staring out of his head.

"W-will you phone the police," he stammered. "I've just found a man hanged in Warren Barn."

"Oh Paddy," I thought. "Your promise…"

The barn was about two miles away in the woods on the edge of the Downs, an isolated spot.

"I went to get some sacks out of the barn," the man explained, as I looked up the number of our nearest police station. "The door was bolted on the inside so I crawled underneath and came up face to face with – him. It was horrible." He shuddered.

"Sit down," I said. "I'll get you a drink."

"No thank you ma'am. If you'll phone, I'll be on my way."

His way, I heard later, lead directly to the Newburgh Arms where he remained knocking back double whiskies until closing time.

The police arrived to collect the body. It was not Paddy but an unsuccessful businessman from Worthing, but we never saw Mr or Mrs Paddy again. We later heard that they had gone to join their son in Australia.

10. A Family of Goldens

When Beauty was two years old I decided to mate her and made enquiries about a suitable stud dog in the vicinity. This is how I met Ann. She was disabled, but one of those intrepid types who achieve more than many who are sound in wind and limb. Ann had one leg in a calliper and walked with a stick but looked after a large bungalow and garden, was a splendid cook, ran a needlework shop in Worthing and bred and showed Golden Retrievers with considerable success.

She had a fine dog at stud called Kim and two bitches, Fay and Jane. The three went to work with her each day, where they behaved with perfect decorum, one sitting quietly by a handmade lace tablecloth, the others in front of a beautiful tapestry.

Ann said she would bring Kim out to the farm one afternoon to make Beauty's acquaintance. When she arrived she had with her all three dogs and as she opened the car door they poured out like a golden avalanche, charged up the drive and through the house in a manner quite unlike that of the docile creatures I had seen amongst the needlework.

Jane and Beauty, startled by this invasion, retreated to their baskets but later when the visitors were settled on the lawn emerged to sit, as they often did, one each end of the step like a pair of bookends.

Kim was big friendly dog and a most suitable sire, bold but gentle, intelligent and affectionate with that extra personality that sets a dog apart from his fellows. He stayed with us several times during the next few years when his own bitches were in season and we became as fond of him as we were of our own dogs.

On this day he and the two bitches looked especially beautiful having been trimmed and groomed for many hours in readiness for a show the following day. Ann had bathed them in a popular brand of ladies' hair shampoo until every golden hair gleamed like spun silk; their ruffs and breechings had been fluffed and brushed until they looked like swansdown.

Their total cleanliness and complete absence of natural smell seemed almost unreal. Jane and Beauty looked like poor Cinderellas beside them.

It was a warm afternoon of early summer and the time passed pleasantly as Ann and I sipped tea and talked on the lawn, the dogs grouped together like the subject of a Landseer painting, the bookends on the step watching them.

When it was time for Nick and Peter to come home from school I got up to get their tea and feed the hens. It was then that we missed Ann's trio. They were no longer to be found in the garden. I looked in the house and on the road while Ann stood and blew her silent dog whistle. At last we saw them running toward us from the field by the house. As they got nearer, then crawled under the fence, Ann and I stared at them in disbelief. Those three silken golden dogs were now wet, slimy, bedraggled and a dull greenish-brown in colour.

The dairy herd had been turned out in that field the previous night and the dogs hadn't missed a cowpat.

We took them into the yard and hosed them down. Jane and Beauty, to whom cow muck was no novelty, had not taken part in this revelry. They followed, now looking clean and groomed by comparison and sat with smug expressions on their faces while we liberally douched the miscreants with water. This removed the worst of the mess but they were still a sorry sight.

"I must take them straight home," Ann sighed. "It will take me all the evening to get them bathed and groomed again."

I put newspapers on the car seats and we piled them in. The natural smell was strong enough now.

"Let me know when Beauty comes into season and I will bring Kim over," said Ann as she left.

One Wednesday a few weeks later she came again with Kim and we put the pair together in a loosebox.

Although Beauty seemed kindly disposed toward him, she decided to play hard to get. When evening came with still no success, Ann consented to leave Kim with us for the next few days and fetch him on the Friday afternoon.

I put them together at reasonable intervals, varying the scene from the stable to the walled garden to the field. Kim's patience and perseverance seemed inexhaustible, but Beauty remained elusive.

When Ann returned I had no good news for her. We had tea and soon it was time for her to go and Kim with her. Sensing my extreme disappointment, she suggested that we should try them once more in the walled garden.

Kim must have realised that it was now or never, for after a few minutes there was a commotion at the far end of the garden, where he had corned her in the rhubarb patch.

The rhubarb never recovered that year but Beauty was successfully mated. It had been a long hot day, Kim drank a whole bucket of water then Ann took him home.

His puppies, when they were born in July, were just like him, light in colour with the exception of one little dark bitch, with black noses and dark eyes.

Beauty was an excellent mother and very possessive, not allowing even me to touch the pups for the first few days, but, when at four weeks I began to wean them, she regarded the proceedings with interest and approval.

There were four dogs which we named Silas, Simon and Sorrel and one larger and stronger than the rest, Stobie and four bitches, Susan, Sarah, Stella and the reddish gold one, Scarlett. They all had their eyes open by the time Nigel came home from school.

There was an afternoon of relaxation when we attended his sports day. I always enjoyed these occasions; the happy end of term atmosphere; strolling in the lovely Italian garden; sitting in the sun, fanned by a breeze off the sea while the watched the fathers versus sons cricket match; eating sandwiches and homemade cakes in the big marquee.

From that day for many weeks after life revolved round the boys and puppies and my chief concern was the filling of stomachs. The small canine ones took three main meals a day with milk and cod liver oil in the afternoons. There were eight little dishes to be washed up four times a day, meat to be minced and mixed with biscuit meal.

Feeding needed strict supervision to see that each pup kept to his own dish. Stobie, given half the chance, would demolish his own then race along the other bowls, swiping the meat out of each. At last I was obliged to shut him in a hen coup with his meal and his yelps of rage could be heard far and wide.

The weather from early June had remained hot and dry and the pups lived in the back garden, spending only the nights in the stable. Shep brought me some hurdles from the Downs and we made a pen on the lawn in the dappled shade of the apple tree. Beauty could climb in and out at will and seemed never to tire of playing with her family.

Jane had no desire to act as grandmother as I have seen some older dogs do. If a puppy advanced in her direction she would beat a hasty retreat to the house.

Because of the many steps between the garden and the stable the pups had to be carried to and fro. This was the first job every morning and the last at night for the boys and myself. At first we tucked one under each arm but after a few weeks when they weighed over twenty pounds each, even one was as much as Nick or Peter could manage. Without a spot of rain on the harvest it was gathered in record time, by the first week in September.

"Let's go down to the sea," Dug suggested on his first free evening.

"What about the pups?" I asked.

"They won't hurt, feed them when you come back. Come on, you haven't been anywhere for weeks."

This was true enough. My days had been fully occupied with preparing meals for my family and Beauty's and all the other essential chores. At night when everyone had gone to bed I sat up filling in yard-long pedigree forms until I was muttering the names of sires and dams in my sleep.

The tide was up at Climping, the sea as flat as oil, a perfect mirror for the crimson rays of the setting sun. A fringe of gentle, fluted wavelets inched over the firm, shining sand. We had the beach to ourselves.

Having put on our swim suits at home we lost no time in stripping off our top clothes and running down to the irresistibly beckoning water; it felt like silk, clear and lukewarm, it caressed us as we swam easily or floated, looking up at the saffron tinted sky flecked with little pink powder puffs of cloud. The long working day, the hungry puppies waiting to be fed and bedded, seemed a world away.

Dug and the boys climbed the high breakwater and dived, the splash as they entered the water the only movement on the calm surface. Even the gulls were immobile, standing rigid in formation on the sand, like cut-out models on matchstick legs.

All too soon we had to leave. Reluctantly we waded out of the water and slowly walked up the beach. Looking back, we saw that the last rays of the sun had disappeared; the vermilion glow was fading; a sudden cool breeze sent ripples across the surface of the water. We had had the best of it.

"What's for supper?" asked Peter. "I'm starving."

At eight weeks the puppies were ready for sale. Stobie had been booked for some time by a businessman who had since gone abroad. I had promised to keep the pup for him until his return in three months.

Susan went first to relatives of Dug's near Bath. A young chemist from Chichester bought Simon. A lady from Pulborough had Stella for her school boy son. Silas and Sarah went to a farm and Sorrel to a large happy family near Horsham. Only little Scarlett was left.

One morning I had an inquiry by telephone from South London. The young woman who wanted a puppy for her children was very enthusiastic but I was not so keen to sell. All the others had gone to country homes and I was reluctant to let Scarlett go to London. I put up my price. The young woman was undeterred. At last it was agreed that she and her family should come down to see the puppy the following Sunday.

Shortly after lunch, an aged Ford Prefect trembled up the drive. A young man and woman got out. No, they had not come to buy the puppy they were only providing transport. A second couple crawled out of the back of the car and unfolded themselves, followed by four children of assorted sizes. It was like watching a seemingly impossible number of rabbits being produced from a hat.

When they were all in the house, I fetched Scarlett. The children, although excited were very gentle with her.

"Have you a garden?" I asked presently.

"Not very big," replied the young Mother. "But there is a large park across the road, she will have plenty of exercise," she added, as if answering my unspoken query.

When she asked for a diet sheet so that she could give her the proper food, I knew Scarlett was going to be all right.

After tea, a rather crowded affair with five of us and eight of them round the trolley and Scarlett leaping on each child in turn, the children had a last romp with her round the lawn, then they all squeezed into the little car, the children piling onto the laps of the grown-ups until we could hear the springs sigh. I put Scarlett in on top and hoped she would not be sick.

The last we heard from them was the Prefect grinding into bottom gear half way up the hill. A few weeks later I heard from the man for whom I was keeping Stobie. He was not returning to this country as expected and was therefore unable to have the dog.

"Let's keep him!" chorused the boys, when I read the letter out at breakfast. "Dad can take him shooting."

"I haven't time to train him," said Dug.

"We'll train him for you," we said.

Finally we persuaded him that Stobie should stay. We had all hated the thought of parting with him and we brushed aside all doubt as to the wisdom of keeping him with the two bitches.

During the winter months we taught him first to 'sit' and 'stay' then to retrieve a stuffed rabbit skin. He got the idea in one. Then to retrieve blind, Nick covering the dog's eyes while Peter hid the dummy in the beech hedge. He found it without difficulty. He learned to walk to heel and became obedient without loosing his gay and sometimes absurdly playful manner. Most days there was some new trick or comical escapade of Stobie's to report to the boys when they came home from school. Like having a lovable child around, he gave a new lift to our family life.

He was nine months old when, late one spring evening a weary middle-aged couple came to the door. Their name was Smith and they had been told that we might have a young dog for sale.

We did have a young dog but he was not for sale we told them, but we asked them in. That morning their much-loved spaniel had died suddenly. They had come away from the house and Mrs Smith would not go home without another dog. But although they had driven all day, from one breeding kennel to another and two lost dogs homes, even up to Battersea, they had not found one that they felt would replace their old friend. At last someone had sent them to Ann and she had told them about Stobie.

"She said you did not want to sell," said Mr Smith. "But we came anyway, it was our last hope."

"May we see him?" begged Mrs Smith.

They sat down heavily, their faces grey with sadness and fatigue. Dug got them a drink while I fetched Stobie, who was having a bedtime romp with Beauty in the walled garden.

He greeted the visitors with his usual exuberance but sat when I told him and gazed at them with his enormous dark eyes, his tail sweeping the floor.

When I saw the look on their faces I knew we had reached the point of no return.

"How much do you want for him?" Mr Smith asked, taking out his chequebook.

I stated my price, knowing that I could have asked much more and he still would not have hesitated.

I could not meet Peter's stare of indignant disbelief or look at Nick, struggling with his tears, but it was done; Stobie went off with his new owners happily and without hesitation. He was fully grown when they brought him back to see us, a grand dog, as tall as Kim but thicker set, affectionate and gentle but determined.

"He is a lovely dog in every way. We are devoted to him," said Mr Smith.

It seemed that Stobie had it pretty good, with so much loving attention and the run of a private beach.

"He's a good dog," said Dug when they had gone, "but headstrong. He'd have been a worry to you, you were wise to sell him."

I knew that he was right.

11. A Year of Events

1960 was an auspicious year for two reasons. In April I passed my driving test and in the summer I became pregnant in my fortieth year.

"It would be nice if I could drive," I had said to Dug after a long time queuing for the bus in Chichester and the two mile walk with heavy shopping from the bus stop to the farm. "It would make shopping so easy and I could take the boys down to the sea in the summer."

In 1940 I had driven an Austin 10 van on a milk round and for the remainder of the war I had kept up my license for the purpose of driving tractors or the odd farm lorry, but since then I had let it lapse. I had never driven in traffic or passed a test.

"You'd never drive on the roads today," Dug told me with conviction. "You'd never pass a test, anyway," he added, as if this were not sufficient deterrent, "and you wouldn't handle the Rover."

I had to concede that our family car, of uncertain age and temperament, needed a specialized technique, so the subject was dropped, but the time came when she quietly but adamantly gave up the ghost and was towed away to be replaced by a reasonably up-to-date Hillman Minx.

I found myself a driving instructor and set off one bleak January day with L-plates tied on with binder twine and my keys on a pig ring as befitted a farm wife, for my first lesson.

I had always imagined that driving a car, like swimming, milking a cow or riding a bicycle was something once learned was never forgotten. In my case this proved to be a fallacy. I know my despairing instructor did not believe that I had ever sat behind a steering wheel of any sort before.

I learned that a country milk round in the war years, with no other hazard than meeting army convoys in narrow lanes, was no preparation for driving in town traffic.

As Dug, perhaps wisely, refused to get in the car when I was at the wheel, getting sufficient practice between lessons was difficult and I relied on various intrepid friends and acquaint-ances such as the neighbouring farmer's son, the postmaster's daughter and Ann when she came to see me, to 'sit in' beside me.

All the same I gradually improved and arrived for my test in April with some measure of confidence which slowly evapo-rated as I waited at the test centre with other candidates and completely disappeared when I at last sat in the driving seat with an inscrutable faced examiner by my side.

"Which way did you go?" Dug asked me afterward.

I could not tell him. From the beginning I had felt as be-wildered as Alice when she fell down the rabbit hole. But I did remember the incident at the level crossing gates, which were closed when we reached them. I recalled many long waits at this crossing for sometimes two long trains, so, feeling rather warm, I decided that this was a good opportunity to take off my jacket. I was half way out of it when a modest little engine chugged into view. With a brief puff of smoke it crossed the road and disappeared behind the station. The gates opened, there was a revving of engines behind me, but I was completely immobilized with one arm stuck fast in my jacket sleeve.

I went through all the contortions of a butterfly emerging from a chrysalis while cars lined up behind, tooting their horns. Then my examiner, who had continued to sit like the sphinx throughout this pantomime offered to help me, just as my arm came free. I put the car into gear and pulled away. The examiner again looked rigidly ahead. Would he fail me for this, I wondered. After all, there was nothing in the Highway Code about not removing articles of clothing at a level crossing.

But what about causing an obstruction while doing so? Never dwell on past mistakes, my instructor had told me, just keep pressing on, which seemed a good maxim not only for a driving test but for life in general.

Everything went very well after that while I drove through the town but as I turned off the main street a man who was walking rather unsteadily near the curb, suddenly fell in the path of the car in an epileptic fit. I did my second emergency stop that day.

The examiner got out and went to help the man but as he bent over him the epileptic scrambled to his feet, threatening with his fists and shouting abuse. Other passers by now came to the rescue and the examiner returned to the car.

"Drive on," he said.

He looked hot and discomforted and completely human. Perhaps that was the reason that I felt perfectly calm and clear headed for the first time that afternoon. I executed a faultless three-point turn in a narrow road with a high camber and a passable reverse round a corner. Questions from the Highway Code were no problem. It had been my chief reading matter for weeks and the boys had drilled me in it well the night before.

All the same I was surprised to learn that I had passed.

When Dug came back I was trying to remove my L-Plates, struggling with the knots in the binder twine. He looked astonished, then pleased, but still declined to let me drive him, preferring to take the examiner's word that I was competent in charge of a vehicle on the Queen's highway.

Later that year I was involved in a slight accident. It was the occasion the boys ever afterward referred to as "the time mother hit the postman."

Now, at no time have I engaged in fisticuffs with any employee of the GPO.

It was like this.

I was driving from church one wet autumn afternoon and as I turned a blind corner into a narrow flint walled lane, I found myself on a collision course with the mail van. The postman and I both applied our brakes but my shoe, slippery with mud from the church path slid off the pedal. Before I could brake again the car had rolled gently into the van. But the impact was sufficient to break the van's radiator, although the Hillman escaped without a scratch.

I was sorry for the postman, who had worked all day with rain pouring down on his head and now had water pouring from his radiator all on account of a fool woman who couldn't stop.

Our local garage proprietor advised us to pay the cost of the damage, which, he said would not be a great deal, rather than lose our 'no claims bonus'.

My replies on the accident form were exemplary in the extreme.

Q. Reason for Journey?

A. Mother's Union Church Service.

Q. Chief Witnesses?

A. Rector and Wife.

It was clearly not a case for the breathalyser.

For eighteen months we heard nothing and just as I thought that everything must be forgotten and forgiven we received a bill from the GPO, vastly in excess of the sum mentioned by our garage man.

It seemed that the Post Office accounts, like the mills of God, 'grind slowly but exceeding sure'.

The news that there was to be an addition to our family was received with mixed reactions; by Dug, with shock and dismay, which took time to dispel. He was eventually consoled by the idea that at last he might get his longed for daughter; by Peter with delight that he would no longer be the youngest. Nick made no comment at the time, but later I heard him say to Peter that he though mummy and daddy had been very foolish. Here they were, nearly grown up (he was twelve) they would look so foolish with a baby sister.

I waited to the end of term to tell Nigel, remembering his remark at the age of four when he was informed about Peter's pending arrival.

"Of course, if you are going to keep on having babies, I shall find somewhere else to live."

I feared that at fifteen, he might spend all the holidays with school friends and I would never see my first born again. But he didn't mind, he said, as long as it was a girl. He couldn't stand any more little brothers.

I hastily reassured him this time it was going to be a girl. Had anyone enquired, I would have said I was as excited as that first time nearly sixteen years before. After all, I had produced offspring in every county in which we had so-journed, Nigel in Oxfordshire, Nick in Somerset and Peter in Hampshire, could I do less for my native county of Sussex?

I have often heard expectant mothers complain that the months of pregnancy drag and I have perhaps been fortunate in never having found this to be so. This time they fairly flew. There was so much to fit in, before our daughter was due to arrive around the Ides of March.

First of all there were those promises to fulfil that I'd made if I passed my test. Although it was not a long hot summer like the previous one, there were pleasant bursts of warm weather when the boys and I made for the beach. As I swam and sunbathed I reflected that those arduous driving lessons had really paid off.

On less clement days, often, I recall, when it was pouring with rain, we went fishing. Again with a car load, plus rods and nets which frequently stuck into the back of my neck, a jar of worms, a tin of mouldy looking ground bait and a carton of maggots, I set off for the Millstream in Arundel.

Nick and Peter had brought the maggots and always very concerned over their condition, insisted on keeping them in the refrigerator, because the cold kept them active. Daily when I was in the midst of cooking they would thrust the squirming mass under my nose with an anxious inquiry.

"Do you think they are all right? They don't look very lively."

They always looked lively enough for me.

Our fishing had altered very little over the years. The shoals of dace still eluded us; Nick still fell in the water and I still got my hook caught in the weeds or a tree root whenever I was invited to cast. Most of the time I sat on a canvas stool under the dripping trees, doling out bait.

In August we all became interested in motor racing, although I had never expected to be. Often when the wind was in the right direction, we could hear cars roaring round the track at Goodwood and I had never felt any desire to get closer to this noisy sport, but when a young cousin of Dug's, Robert Duggan, spent a few days with us while driving in the races, I caught the enthusiasm along with the rest of the family and went along to watch.

He was driving his own Morgan Plus Four and the boys were in it and all over it like Martians over an earth satellite, directly he arrived at the house. Next day with great excitement they went with us to see him in the pits before the start of his race. Then we took our seats in the stands.

Goodwood had never been as lucky for Rob as Silverstone, but that day he was second away at the Le Mans start and was in a good position at the end of the first lap. He flashed past the stands and for a while was lost to our view.

Then we heard the commentator exclaim, "Duggan has lost a wheel, and he's spinning across the track on three wheels and back again."

We held our breath; the boys were rigid with suspense. Then ... "He's alright. He's on the grass verge. He's pulled up and he's still upright."

When we saw Rob again back in the paddock he was as calm and unmoved as if spinning across the track amid on-coming cars was a daily occurrence. He shrugged philosophically. Goodwood was never lucky for him.

Our next visitor was Mrs Chalcraft. She looked askance when we told her about the baby. I think she considered that another horse would have been more practical.

Some of my most precious moments of relaxation that summer were spent in my greenhouse. There was nothing more soothing than half-an-hour spent pottering in it, picking the side shoots off tomatoes, the male flowers off the cucumbers, pricking out seedlings or just admiring the results.

This refuge was cunningly positioned, ostensibly for the maximum sun and shelter, at the far end of the walled garden and once I inside I could not hear the telephone, various odd bods connected with the farm banging on the back door or members of the family shouting 'Mother'.

I bought it out of the proceeds from Beauty's puppies. Dug said that one would think, judging from the number of things I bought with that profit, she must have had at least two-dozen pups. Be that as it may, I got my greenhouse, which the family at first regarded tolerantly as something to keep me amused, but were subsequently amazed to learn that the succulent tomatoes and cucumbers they were eating were the product of my new 'toy'.

I also walked a good deal with the dogs knowing that it would be some before I could again roam freely through woods and over fields unhampered by pram or pushchair. Those little seats in which parents can now carry their off-spring strapped to their backs, were not then invented and my attempts to make a sling such as Indian women use had never been successful.

One day returning home through the woods the dogs and I had a strange experience with a rat. Beauty had disappeared ahead and was foraging in some bushes where she put up a rat, which ran along the path towards me. Either mistaking the scent or chasing after something else, Beauty took off in another direction. As the rat got closer I sidestepped, expecting it to run past or scamper off into the wood. Instead it half turned and leapt straight at me.

"Beauty!" I shrieked. She immediately appeared on the bank to my left and sprang simultaneously, killing the rat in mid-air.

It was unusual behaviour on the part of the rodent, which as a rule only attack when cornered. Perhaps thinking it to be pursued and finding me in its path brought on the aggression. Whatever the reason I was grateful to Beauty for her split second timing.

It was not until the autumn that I started thinking about preparations for the new baby. Having long supposed that our family was complete, we had given away all baby equipment and must now start afresh.

Our pram we had given to a family in the village some years before, for collecting wood. Although they kindly offered to let us have it back we said we would manage. Eventually we borrowed a deep safe, comfortable carriage. A trifle aged, it had one peculiarity, a decided list to starboard, so that if one let go of the handle when pushing, it would come round full circle. As I only used it in the garden this was of no importance.

Nick and Peter were enthusiastic helpmates. Through the winter months they amused themselves painting the nursery furniture and took a great interest in the baby's 'trousseau'.

Peter grew impatient for the new arrival and having watched a Western in which a woman gave birth to her child immediately after being thrown from a wagon, he suggested that I might hurl myself from the Land Rover or kindly offered to put down a mattress for me if I would prefer to jump from the bedroom window. I turned down both these suggestions. If I could wait until March, I said, then so could he.

12. Now We Are Six

The third month of the year came in cold and grey. I set out on my daily walks with the dogs against a biting North-Easter. Our 'daughter' was due on the fourteenth of the month about the time that the first ewe was due to lamb. I heard that Shep had a wager with the herdsman that this 'yo' would be the first to give birth.

A couple of days before this time I was about a mile or so from home on the top Downs when I had the first indication that the arrival of our fourth was imminent. I decided that it would be prudent to retrace my steps rather than continue on my usual round, but I stopped to talk to Shep and Ethel.

Ethel, in spite of the army greatcoat that enveloped her, shivered with cold and Shep blinked at me with eyes red rimmed and watering in the wind. There was a bright glow to his pipe and to the end of his nose.

"Aah, tis a lazy wind," he declared. "Don't goo round ee, do goo straight through!"

I enquired after the ewes, as yet there was no sign of lambs.

I could have told him he was about to lose his ten bob, but it could be a false alarm.

It wasn't, but there seemed no urgency, so when after tea Dug said he was going to shoot pigeons on Nore Hill I thought it a good idea. While he was gone I packed a case.

Then we all sat down to watch the Saturday evening television programmes.

It wasn't until the middle of the western that I began to wonder if I waited to see how the hero escaped from the Indians, whether I would make it to the Maternity Home some ten miles away.

I looked at the boys' tense, wide-eyed concentration. It seemed a pity to break up the family viewing.

In the end I saw the film through to a satisfactory conclusion and we arrived at the home with time to spare.

It was another boy.

Somehow I'd known all the time it would be. We called him Jonathan.

Dug swallowed his disappointment nobly and Nigel said "Oh well, I'll be at school most of the time anyway".

Like two of his brothers, Jon was born lean, dark and hairy. Only Nicholas had surprised us with the blond hair and blue eyes of his paternal grandmother. His black hair was of such length and thickness that a friend who brought him a present of a brush and comb set asked if I would like her to change if for a hound glove.

The first ewe did not lamb until four days later but she still stole my thunder by having it delivered by caesarean section. The weather changed the weekend Jon was born and during the ten days that I spent between four walls warm sunshine transformed the countryside. When Dug drove us home the gardens were gay with daffodils, forsythia and flowering trees.

At Court Hill the daffodils were out on the bank and small tortoiseshell butterflies fluttered on the aubrietia.

I gave Jon his first feed every morning to the accompaniment of the dawn chorus and when I put him to bed at night, pipestrells flitted outside the window. Spring was a good time to bring home a new baby.

Nick and Peter welcomed their new brother without reservation. Nick, far from feeling foolish with a baby, exhibited him proudly to all his pals, turning back the hem of his gown to show off his tiny feet. As for the gang, he was the greatest thing to hit them since transistor radios. They took him into their midst with an enthusiasm which he miraculously survived. Often when I looked into the pram he was missing and I would find him with the gang in the yard or some part of the garden perhaps being nursed by Tom while the others clustered round. When he thought of something else to do, he would toss Jon like a rugby ball into another pair of arms. They carried him round the fields and when camping, smuggled him into the tent.

One day when I went to fetch him for his feed, both pram and Jon were missing.

"Looking for your baby?" asked one of the men passing the gate on his way back from dinner. "The gang have got him down by the village pond."

Except during school hours there was always an admiring circle round the pram. When I bathed and changed him in the big bathroom they gathered round, one playing a transistor radio, another a mouth organ, perhaps one a guitar. When I left him to fetch a garment from the airing cupboard I had to elbow my way through the throng to get at my baby again.

Nick was chief nursemaid and usually displayed very good sense but one day he startled me. He had always surpassed his brothers and other members of the gang in climbing. Often I had seen his blond head appear at the tipperty-top of the tall cypress tree. One day when he was seven the others shut him in the loft over the well house and he escaped by climbing through the tiny window, along the side of the office roof and onto the garden wall, while Dug and I stood in the yard holding our breath. It had long been a habit when he wanted to keep something to himself, be it a book or pet animal or something edible, to scramble with it to the top of a tree. Even so, I was amazed to see him sitting on a high branch of the tallest apple tree nursing his three-week-old brother.

In spite of all this unorthodox treatment Jon thrived. His days and nights were trouble free and he put on weight so quickly that Dug said if he could breed a pig with a conversion rate like Jon's he would make a fortune.

His christening in May was a happy family gathering. Only Dug was missing. Usually very healthy, he chose this inopportune time to go down with flu and was unable to attend the ceremony. Even so, I must have been in a more collected frame of mind than at Nick's christening twelve years before. On that occasion I had dressed him in the long embroidered gown and petticoat and laid him carefully in the centre of our big bed while I got ready.

The church was only a short distance from our cottage so we walked along the lane, chatting happily with our guests in the warm June sunshine. Suddenly the godmother stopped.

"Who's got the baby?" she asked.

We all looked round. No-one had the baby.

I had left him in all his finery, lying on the bed.

During our years at Court Hill there were certain pressure periods when I could have done with an extra pair of hands, at least two extra pairs of feet and twelve more hours in the day. I entered one now with the start of the summer holidays.

Nigel worked on the farm but the other two and the gang enjoyed the good life, attending to their camps, cycling to the beach or nearest swimming pool or, on wet days to the cinema in the town. I felt slightly envious.

My own days were completely housebound, divided between kitchen and bathroom, while I fed and generally attended to Jon. They began at 5am with his first feed and ended at 11pm when I put him down for the night. In between the round of baby routine was the washing and cooking, the endless sessions of staggered mealtimes. When my energy reached rock bottom there was one thing that kept me ticking over. It was the constant flow of pop music. For some years now the current 'hits' being played at full volume on some boy or other's transistor had continuously bombarded my ears. I had to become a pop fan or perish. On the basis of 'If you can't beat them, join them' I enjoyed the reputation of being the most 'with it' mum on the subject of groups and the 'charts'.

"I thought classical music was your style," said Dug, a little askance at my new enthusiasm.

"It still is," I replied, "But for it one must have time to relax and listen, I can listen to pop while I work and I don't feel guilty if I clatter a saucepan in the middle of it."

It was a vintage year for pop singers: Gene Pitney, Roy Orbison, Elvis Presley, Roger Miller – his rendering of "King of the Road" still takes me back to that happy, hectic summer –

and of course the Beatles. When I was exhausted, the 'beat' of one of their numbers would pick me up from the floor and set me ticking round the kitchen again. For that alone they were on my personal honours list.

Jonathan had his own preference in Kenneth McKellar. If he would not settle down in the evening I had only to nurse him for while in the rocking chair to the accompaniment of 'My Love is Like a Red, Red Rose' and he would be sound asleep.

Unfortunately, someone left the record on top of the radio and it became corrugated in the heat. I think it ended up as a plant container.

One evening the piano arrived. It was a genuine honky-tonk of uncertain age. The first time I set eyes on it, it was being pushed up the drive on a sack truck by three panting, red-faced boys. By dint of much pushing, puffing and heaving they had it through the front door and into the dining room where they made a space for it along one wall. From that time onward the jangling tones struck from it by various inexpert fingers added to the general cacophony.

"How does your baby sleep through this?" asked a caller one evening, when David, Richard and Nick were listening to a pop programme on television, Tom was sitting on the stairs with a different selection blaring from his transistor and the remainder of the gang were taking turns on the piano.

"He sleeps in self-defence," I answered. "He knows he can't compete with it."

In August, Rob came to visit us again. He had sold the Morgan and was driving a Lotus Elan for a sponsor. He brought with him a tall, auburn-haired girl named Margaret, whom I found very pleasant and helpful.

As there was no corn cutting that weekend Dug said he would look after Jonathan while I took the other boys to see Rob race. When we arrived at Goodwood we found the Elan and Margaret in the paddock waiting for Rob to appear. I imagined she must spend a lot of time waiting in the wings, dishing out tea from a thermos or by the side of the track timing Rob's laps.

The Elan had lost a sidelight when Rob had spun off the track at practice that morning; the Goodwood jinx was still on. When Rob's race came up we found a good vantage point, the boys as usual hopping with excitement. Rob's chief opposition on this day was his friend and often co-driver, Mike.

When the cars passed us on the first lap several were jockeying for position but Mike and Rob were already out in front. There was a patch of oil on the track and three cars span madly but Mike and Rob were clear. By the third lap it was apparent that the race was between the two friends.

Rob was in second place but as usual he gained ground through the chicane.

"Half an inch to spare is too much," he always said, and was only sixteen seconds behind Mike.

They were now approaching the bend where Rob had lost the track at practice and perhaps using too much caution he lost ground. In the last lap he fairly shot through the chicane but behind Mike, who was first to pass the chequered flag.

When Rob and Margaret came to see us the following summer they were on their honeymoon, spending one night with us on their way to Holland, having spent the first two days after the wedding at Silverstone where Rob was racing. I reflected that Margaret and I something in common. If the farmer's wife came second to the farm, the racing driver's came second to the car.

Soon afterwards Formula 1 racing ceased at Goodwood. On my next visit there, Peter would drive me round the track – but that was many years ahead.

When harvest was over Dug and I took a day off and leaving the boys to fend for themselves, took Jonathan and drove into Hampshire to see Mrs Chalcraft.

We found her in her usual good spirits. Her Dexters were looking well; she had a pair of young Spinger Spaniels, which were a delight and she was full of enthusiasm about a new way of growing tomatoes.

Our conversation, as ever, was of farming and horses. Horses were represented in the photographs, drawings and oil paintings which covered the walls of the large, gentile, shabby rooms. Evocative of a past era they opened out onto long smooth lawns fringed with colourful borders of old-fashioned flowers.

Unfortunately, Jonathan, usually such a good baby, was restless that day and howled lustily for much of the afternoon, a performance that did nothing to convince our friend that we had done a clever thing in adding to our family at our time of life.

It was the last time we ever saw her. The following winter we heard that she had died, suddenly and peacefully in her eighty-fifth year.

If I had expected Jon's arrival to be a nine days wonder to the gang I was wrong. During the next summer they followed his development with as keen an interest as they had the first. They had all acquired cameras and frequently gathered round him like a gang of press photographers.

They proudly showed me the resulting snapshots of Jon in his cot, Jon in the bath or paddling pool, in his pushchair against a background of lambs or piglets, perched on their bicycle saddles or simply being held by one boy or another.

This second summer was eased for me by the arrival of Helen, a friend's eighteen-year-old daughter who came to help me while waiting to start her career as assistant matron of a boy's school. Our home seemed good pre-experience. Helen was fond of children and took over most of Jon's routine. It was she who taught him to take his first steps on the big kitchen table. Although rather more sophisticated than our harum-scarum she entered well into the fun and liveliness of our household.

It was the year of the Twist. There had been Rock 'n' Roll, the Jive, we'd been through them all, but now they were superseded by this new delight. To the invitation of a certain Mr Chubby Checker on television, radio and record player, our gang with the rest of the nation's teenagers, twisted again and again. They twisted in the kitchen, in the pantry and in the hall. They twisted in the lounge, their heels gouging out the last threads of our already well-worn carpet; they twisted in the bathroom and on the landing.

Helen, who was tall and lissom, twisted beautifully and Nick had found another outlet for his energy and agility.

I never tired of watching them. The dance completely fascinated me but I was warned that one should not attempt it if over the age of forty.

13.   The Money Spinners

That Spring Nicholas launched the first of his moneymaking schemes. "You know Roy...?" he asked one evening when I was cooking supper. I had noticed that in childhood, throughout adolescences and after, the boys divulged their most private confidences and most profound items of information to me at the kitchen stove. Perhaps because that is where I was most often to be found.

"You know Roy?"

I knew Roy, he had haunted the house daily for several months and he was not the sort of boy one could overlook. Tall and lanky, he had bright ginger hair that reached to his shoulders and keen blue eyes that shone from behind large horn-rimmed spectacles. His clothes too, could scarcely escape notice, for he favoured bright pink and purple jeans with gaily contrasting shirts. His outfit was invariably topped by a brown pork pie hat, which was decorated by a sort of mobile natural history museum; a hare's foot, a jay's wing, a pheasant's tail feathers, the claw of some other bird. Roy had a ready wit, a brain that earned him a remarkable number of 'O' Levels with little apparent efforts and a persuasive manner that could charm the birds off the trees.

"Well, we thought of breeding rabbits. You can make a lot of money from the meat. Do you mind if we build some hutches at the bottom of the garden?"

I would, I said, prefer the garden uncluttered by rabbit hutches, but I couldn't stand in the way of a paying concern.

Nick carefully pointed out that I was not expected to get involved, which was just as well, because I had no such intention, and I didn't … at least, not for a while.

Nick got to work with his usual enthusiasm and soon an untidy pile mounted at the end of the walled garden; empty boxes from the grocer, oddments of timber from the woods, planks and pieces of corrugated iron from around the farm, until it resembled Steptoe's yard. Then he began to build platforms and make hutches. I couldn't quite see where Roy came in, except that he turned up most evenings to see how things were going and occasionally he held a few nails.

I tentatively asked Nick, what his partner was doing to help.

"Oh, he's paying for the first rabbits," was the reply.

Well, I supposed that if you were financing a venture, you didn't expect to do all the spadework as well.

At last the premises were ready and five pregnant does, a buck and two maiden does were installed.

We all went to look at them and admired their sleek coats and healthy appearance and stroked their soft waffling noses.

Nick got up early every morning to feed them before going to school. He fed them again as soon as he came home. There was green stuff to be collected and lots of little dishes to be filled with water and meal.

I watched it all, so far blissfully uninvolved.

Then, one teatime there was a telephone call from Nick. He was detained at school for drama practice and could I please feed the rabbits. Only then did I realize what a chore it was. I sent Peter to collect the green stuff.

Roy still came most days and gave advice and brought a mass of literature which lay scattered about the kitchen and pantry; leaflets entitled, 'Rabbits for Meat', 'Make Money from Rabbits', etc… It all looked very impressive.

After a week the first doe had a litter of five young. Nick glowed with importance like a proud father. Roy came along to offer his congratulations and advice on the rearing.

Soon every hutch was seething with young. Nick's work increased daily but so did his enthusiasm. Roy admitted that he was doing very well. About that time Tony joined the gang. Newly returned from Kenya with his parents he added a fresh dash of interest and colour to the group. Fair-haired and robust, he had that extra 'savoire faire' so often found in a child brought up abroad.

Tony's father, just retired from a government post, was 'great' the boys declared and 'could do just about everything'. Certainly, I thought, he could have been the originator of 'do it yourself'. At that time he was engaged in transforming a dark, rambling old house into a home of charm and elegance.

Tony's mum, the gang said, was a terrific sport, she was very funny and her cooking was out of this world.

There was a lot to do at Tony's place, a pony to ride, new ground to explore and a host of interesting things going on. Nick spent a lot of time there. Roy joined him and gradually all the gang deserted for Tony's place.

It was very quiet with just Beauty and Jon left at home. He was bored and disgruntled without the gang to amuse him. I wondered what Tony's mum thought of the invasion. Oh well, no doubt they'd be back…

Meanwhile the rabbits were doing well. Fed on pig meal they grew to enormous size. Eight were ready for the butcher and Roy, who handled the business side, had contacted a man in Chichester who would give him a good price. There was just one problem. Who was going to kill them?

Roy firmly declined. That sort of thing was definitely not his department; Nick, who had reared them with such loving care, was quite averse to the idea. In instances of necessity I had killed hens, rats and mice, but at tame rabbits I drew the line. Besides, they were not only big but strong; dispatching them would need an expert hand. Dug, who had no objection to killing rabbits with a shotgun, merely shook his head and drove away in the Land Rover when asked to help out in this predicament.

It began to look as if, after all, the rabbits would have to die of old age when Peter had a brainwave.

"What about George?" he asked.

"Of course!" we all exclaimed. "George the rabbit catcher, just the right person."

George was approached and agreed to act as executioner. One Saturday morning Roy and Nick rode into Chichester with eight heavy carcasses hanging from their handle bars and returned jubilant with the first proceeds of their hard work and enterprise.

The does by now had more young growing fast. The thing was a going concern, but as so often happened with breeding animals of any sort, it is just when things are going really well that ill-luck strikes.

One evening Roy turned up with a pretty little black doe and her litter. He had got them cheap from a friend, he said.

Nick eyed them without joy. They didn't want to over stock, he said, and it meant knocking up another hutch to house them. With a little help from Roy he had one ready for the new family that night. I went out to see them. The doe didn't look very lively. Her eyes were dull and she ate only a little of the food Nick put in for her.

"She's not well," he remarked.

"Oh, it's just moving her with her young," Roy replied complacently. "She'll buck up".

But she didn't. Next morning she was hunched up and not eating at all. The young looked very weak. Throughout the day she got worse. I tried to revive her with a little brandy but when Nick came home from school she and two of the babies were dead.

This was when I got involved.

"Mummy," said Nick. "Do you remember you used to do autopsies on the guinea pigs when they died?"

I remembered.

"Well, could you do one on the doe? We must know what she died of."

It was a messy business. Nick and Roy stood yards away from me with averted faces and held noses and checked the symptoms I described from a booklet called 'Rabbit Diseases'. It was as we had feared, the dreaded *coccidiosis*, from the *streptococcus* virus that causes this disease in rabbits and poultry, mastitis in cows and sore throats in humans.

We burned her remains and those of her young and disinfected their hutch. A few days later another doe had a litter and afterwards seemed not too well. She died the following day and her litter was too young to save. The next week two more were lost. Nick and Roy looked glum. Then for a while all seemed well. Two does reared healthy litters. Roy replaced two of the lost does and things looked bright again. Soon there would be another consignment for the butcher.

Then one morning Nick found a buck dead in his hutch. The plague had struck again. When the new does fell sick the boys decided to cut their losses. They sold the remaining healthy rabbits and gave up the enterprise.

Nick was not one to cry over dead rabbits. Very soon he had another idea.

One morning I entered the larder at my usual breakneck speed (my frequent dashes from kitchen stove to larder had all the economic timing of a four minute mile) and came face to face with a large Tudor Lady, who stared at me haughtily from a canvas roughly five foot by three. She had auburn hair, a creamy pallor and an ample bosom constrained by a bodice of rich velvet and looked so regal I wondered if I should drop her a curtsey.

I returned to the stove seething with curiosity just as Nick came down to breakfast.

"Who's she?" I asked, nodding in the direction of the canvas.

"Well, you know Tony?" was the ambiguous reply.

"Yes. Is she one of his ancestors?"

"Of course not! He and I are going to make a lot of money out of old oil paintings. We'll buy them cheap from the antique market and Tony's father has taught us how to restore them. We might even find an original masterpiece."

"Hmm…" I murmured doubtfully. "Well, you never know your luck."

"She only cost £1," he said, indicating the lady in the larder.

Very cheap, I thought, for all that air of aristocracy.

That very evening Nick got to work on the Tudor Lady. There was a lot to be done. The canvas was dirty and worn in places and there was a small hole, remarkably like one made by an air gun pellet, right in the middle of her bosom. The larder soon reeked of linseed oil, meths and turpentine. There were tubes of paint for touching up and an assortment of messy little rags lying about the bench.

Tony was conspicuously absent from the scene of operations. Perhaps he was buying the paintings.

Every Saturday morning he and Nick visited the antique market in the town and returned with at least one dirty old painting. They were duly cleaned up then disappeared, sold, I hoped for the anticipated profit. So far they had found nothing resembling a masterpiece.

The Tudor Lady remained, her bullet hole neatly mended, her colouring restored. By now we were on a nodding acquaintance and she looked quite friendly. I was going to miss her when she went.

Then one day she was gone. There was an emptiness about the larder.

"Sold the Lady?" I asked Nick.

"Not yet. A dealer offered Tony £15 for her, but he said if he offered that much it must be worth a lot more, so he turned it down."

As the venture had, up to date, barely broken even, I thought a profit of £14 couldn't be bad, but then I am no businesswoman… I soon learned that the Lady had taken up new quarters in Tony's father's garage.

In time the enterprise went the way all such enterprises. The tins and tubes and bottles, the messy rags and the remaining canvases were cleared from the larder and the attendant aroma faded, but before I had time to rejoice in the new tidiness they were replaced by an astonishing collection of bicycle parts.

The idea, Nick said, was to build one complete new bike out of three old ones. This time there was no partnership; he was going it alone. This 'new bikes for old' business caused an even greater state of chaos in the larder than the restoration of paintings.

The bench was strewn with nuts and bolts, brake blocks, chains, handlebars and other paraphernalia; inner tubes and tyres were draped from nails and hooks; wheels and frames stood on the floor; I had to climb through the fork of one to get to the bread bin.

One might think that on a farm there would be places other than the larder for this sort of hobby, but when I tentatively raised the point I was given a list of unanswerable reasons why this was not so.

The stables and garage were too far away from the house and too full of farm equipment, the well house was too cold and draughty in winter and anyway, that and the wash house were the approach to Dug's office and he was very particular about keeping it clean and tidy, so the larder really was the only available place. The expanse of brick floor, the many wide shelves and the long bench made it ideal.

I was allowed the marble slab in the window and three shelves for eggs, stores and preserves so what had I to complain about?

For many weeks Nick spent all his spare time taking to pieces and reassembling the bikes but although the pile of spare parts grew and to them was added little pots of paint ready for the one complete bike that was to rise like a Phoenix from the ashes, it never seemed any nearer completion.

Then, suddenly it was all abandoned, for Nick, along with Roy, Tony and Peter had hit on the greatest idea of all...

14.   The Cellar Club

"Is it alright if we have a club in the cellar?" Peter asked one day. I didn't see why not. It would certainly be easier on the stair carpet than the one they'd had in the attic.

"It's in a shocking state though," I told them. No one had done anything in the cellar except dump unwanted articles of all types since my abortive attempt to grow mushrooms there many moons ago. The soil was still there and the whole place was damp and a sanctuary for creepy crawlies.

"Leave it to us," said Peter.

I did, and a few days later a small working party got busy. It consisted of Nick, Tom and Roy who were to be the club proprietors, assisted by Keith, a newcomer to the gang, Peter and Tony.

The cellar became a hive of activity. The only way into it, except down a shaft from the garden was through the kitchen and pantry. All the first morning boys filed pasted me with buckets, baskets and a large zinc bath, returning with them filled with dirt and debris, the remains of my mushroom bed and every conceivable sort of junk. After several of these excursions they moved in with brooms, while the sort of dust cloud that might have followed a wagon train, rolled up the stairs and into the pantry. Spiders that had thought they were settled for life in their corners ran helter-skelter from the onslaught.

Then the buckets went down again, this time filled with hot soapy water. Every brush and mop in the house and on the farm was commandeered. The walls were washed and diligent scrubbing revealed a red brick floor. The steps were really clean and the whole place had a fresher smell.

The next day the boys arrived with pots of paint. Red went on one wall, green on another, white for the alcove where, later in the week, Tom and Keith built a bar with a wide mirror, found among the rubbish, fitted behind it. The wooden edges of the steps were also painted white for here the light was dim.

"Have you got any chairs and things you don't want?" Nick and Peter asked one morning.

I said that short of taking the dining chairs and the ones in the kitchen that we used daily they were welcome to what they could find. They produced two from the attic, one with a broken leg that was put in the larder for repair, two more from the top bedrooms, an armchair from Nigel's room and an old card table.

Other parents were persuaded to part with their surplus furniture and during the next few days I opened the back door to Tom with another chair, Keith with a table, Roy and Nick with a long form, all of which were transported the several miles from their various homes either on or between bicycles.

When all was ready we were officially invited to view the transformation. The sides of the stairway were covered with huge, gay posters; almost life size pictures of the Rolling Stones decorated the cellar walls, but the 'piece-de-resistance' was the bar. Through the little archway the alcove had a rosy glow from the red-shaded lamps. Colourful labels and beer mats added to the decor.

Roy had procured an air extractor, which was fitted in the blocked up shaft and run off a reading lamp in our sitting room. An adventurous cable passed out of our window, along the rose bed, over the wall and down the shaft.

Dug and I looked askance at this and some of the other electrical devices and called in the local electrician to vet them. He assured us that the boys knew what they were doing and all was safe.

Tom now spent hours in the local dining room recording pop music onto tape for use on club nights, playing it back for the approval of the others, then playing it again each night in the cellar. There wasn't much in the pop charts I didn't know by heart. Meanwhile Nick fixed amplifiers in each corner of the cellar.

The first Saturday night club was a huge success, although Dug and I sitting in the room above were somewhat shattered by the end of the evening. Against the hum of the fan, the over amplified music, the voices and laughter, our television barely held its own when turned to full volume.

Roy's parents, who ran a transports café, provided crates of Coca-cola – again transported by Roy and Nick between two bicycles – and cartons of potato crisps. A charge was made for admission and profits were to be ploughed back into the club in the form of foam cushioning for the seats, Formica for the top of the bar and a plastic strip curtain for the entrance.

During the winter the numbers increased until one night a record twenty-nine crowded into the cellar. They came from a wide radius on a fleet of motorbikes and scooters, which were parked in the yard. The boys shed their gear; leather jackets, parkas and helmets in the washhouse while the girls used the pantry as a cloakroom.

As a direct result of so much 'coke' consumed during the evening we frequently met a queue of mini-skirted girls on the stairs on their way to the loo. They smiled at us shyly when we emerged from the sitting room, while in the kitchen I was politely acknowledged by a stream of young men on a similar errand to the yard.

It was the beginning of the Easter holidays when Peter announced that the club was going to have its own pop group. The next morning two 'new' boys whose names I learned were Baz and Anthony, arrived with electric guitars. Nick and Peter conducted them to the cellar and after a while some strumming ensued. Then Roy arrived with a small drum. Practice on these instruments continued throughout the day, while rain fell steadily outside. After tea there was another knock at the back door. In the washhouse stood two girls, water streaming from their bedraggled hair and limp raincoats. Between them they bore a large object wrapped in sacking.

"It's a drum," they explained, beginning to take off the wet sacks. It was a huge, bass drum with a cymbal on top.

While I took their wet coats and put them to dry and fetched them towels for their hair, I learned that they brought their outsize package on the bus from the town and carried it through the teaming rain the mile-and-a-half from the bus stop. There had been much speculation in the village as to the destination of this mysterious load.

"All we need now," said Tom, "is a lead guitarist."

By the following week, one had been enticed from a disbanded group in Bognor. He was called Diz.

The group was complete and thinking of a name for themselves. I suggested 'The Spyders' out of courtesy for the former occupants of the cellar but for some obscure reason they decided on 'The Sawdust Caesars'.

Every morning promptly at nine o'clock, Diz, Bas and Antony, the guitarists, turned up at the back door. They were joined in the cellar by Roy and Tom on the drums, Nick who supervised the electrical equipment, Keith, Tony and Peter who just listened and made suggestions.

Practice went on continuously for hours, stopping only at mealtimes. I was concerned about these boys being incarcerated in the cellar throughout the lovely spring days and expected them to come up looking as white as maggots but they remained a very healthy colour.

Because of the dampness of the cellar, the instruments, when not in use, were kept in the dining room, the big drum occupying one window.

This, especially the cymbal, fascinated Jonathan, now aged three. Whenever he could creep un-noticed into the dining room, he would seize the drum stick, climb onto a chair and bash the metal disc with a force that threatened to send it spinning through the glass pane and sent reverberations echoing through the house, while he leaned back blinking with surprise and delight. With so much practice the group improved rapidly and soon sounded most accomplished. Very soon they embarked on their first professional engagement. I never did discover the outcome. Their response to my inquiries as to how they got on being evasive. As far as I know it was their only public appearance.

I was relieved after the holidays to find that practice was restricted to two evenings a week and at weekends the gang sort outdoor occupations.

One morning Nick called me out to the yard.

"Come and look at this," he said.

I found Tony and Roy tinkering with an extraordinary contraption. It was basically a tandem, the frame of which had been retrieved from a scrap heap and built up from Nick's collection of spare parts (so they came in useful after all). It had then been painted in bright hues of red, green and yellow.

On one side of this machine the shaky looking chassis of a motor sidecar was attached with pieces of wire. On this rested three rough planks.

"What on earth are you going to do with that?" I asked.

"We're going camping with it at Whitsun," was the reply, "Me, Tony and Roy."

"Are one of you going to ride on the planks?"

"No, that's for taking all our equipment, Roy is going to ride his bike."

I didn't like the sound of it. The contraption looked most unsafe to take on the road at any time, but especially on a bank holiday weekend. Tony's mother wasn't happy either when we conferred on the telephone. Then Dug, Tony's father and Roy's got together and put their combined feet down. The boys were not to go.

This of course was before Roy talked to them. Roy could always be relied on to talk his way in or out of anything. With the blandness of a slick car salesman disposing of a dubious vehicle, he assured us that the machine was legal, roadworthy and as safe as houses, all the things we had said it was not.

The opposition removed the three got to work making their preparations. Roy's parents provided most of the food, enough catering size tins of meat, fish, baked beans and fruit to sustain a small army. Then there was the tent, sleeping bags, cooking utensils and other camping paraphernalia, plus fishing tackle and a radio.

It looked as if nothing less than a van would be needed to transport the mounting pile of equipment, but by sheer force of determination it was all packed into one pregnant load on the planks, until the chassis wheels leaned outward under the weight and the thing in motion weaved to and fro in an inebriated fashion.

I suggested that they should take one of the homing pigeons in case they got stranded miles from a telephone, but they said they couldn't fit on another thing and I had to agree with that.

Their destination was a site by the river at Fittleworth, a village about seven miles away. As a compromise to their uneasy parents they agreed to camp for only one night and finally got away on the Sunday afternoon. The traffic, they said would not be so heavy then as most holidaymakers would have already got where they were going.

As it has never been our habit to go out on a bank holiday we sat in the garden the following afternoon, wondering where the trio were, while not admitting that we were actually worried. They should be well on the way home by now if they were going to miss the returning holiday traffic. When after tea there was still no sign of them, Dug and I looked at each other.

"Come for a drive?" he asked.

We were not exactly going to look for the boys, but it so happened that we drove in the direction of Flittleworth. We did not see them on the road although three boys, one with red hair and pink jeans and a tandem extraordinaire would be impossible to miss, nor were they at the camping site.

It was on the way home, on a steep hill near Arundel that we overtook the sorry crew. Nick and Tony wore ragged, faded jeans rolled up at the bottom and t-shirts in shades of red and purple that could only belong to Roy. They contrasted with their complexions, which, either from fatigue or rich living were a curious greyish-green. Their hair hung lank and dis-coloured, Roy's straggling about his shoulders.

We waved them into a lay-by and pulled up beside them.

"What happened?" asked Dug.

"We fell in the river," replied Tony.

"All of you?"

"Well, you see…" Nick explained. "We were washing up before breakfast…"

"We hadn't felt like doing it after supper," Roy interposed.

"Nick was washing the plates in the river," Tony took up the tale. "At the bottom of a steep bank and Roy and I were hang-ing on to him."

"Then he slipped," Roy went on, "and pulled us both over his head. We all landed in mid stream."

"And the drying up cloth," Nick added.

Miraculously Roy had retained his spectacles and his hat.

"Then there was this pack of hounds," said Nick.

Dug and I exchanged glances. Apparently we still hadn't heard it all. We settled down to listen. It was a story that made the adventures of Jerome K. Jerome's *Three Men in a Boat* seem tame.

"We'd only remembered to bring one towel," added Tony. "And these beagles came over the bridge into the field and invaded our camp."

"We heard them knocking over our pots and pans and gobbling up our breakfast," said Nick. "But we couldn't do anything about it because we were all starkers."

"They ate two pound of sausages, two tins full of beans and a whole cut loaf," declared Roy indignantly.

"They'd all been in the river and were dripping wet. Then they burst into the tent and trampled all over our dry clothes, all except Roy's."

This explained the general turnout in Roy's garments.

There had been some trouble with the sidecar too. As we drove up beside it we noticed a strange undulating movement of the outer wheel. A large bulge of insulating tape on the tyre, we now saw, caused this.

"Oh that…" Roy said in response to our query. "Well, we were going down Bury Hill, when there was this smell of burning rubber. With all the weight on the trailer the overheated rim had singed the inner tube."

"We hadn't any patches because we'd used the last on our fifth puncture," Tony explained.

"Sixth," Nick amended.

"Sixth, but we had this reel of tape, so we just bound it round over the hole, it bumps a bit."

This seemed to me an understatement.

"You haven't been stopped?" Dug asked.

"Well, there were two policemen, they looked at the machine, the first one didn't seem to know what make of it," said Tony.

We were not surprised. It was enough to puzzle the most sagacious member of the West Sussex Constabulary.

"The other one was a bit severe. We let Roy talk to him. In the end he just told us to get off the road as soon as possible."

We firmly endorsed that advise. The traffic passing on our right was getting closer and faster as we talked. A mile or so further on, just off the main road was an expanse of grass and furze backed by woodland. It was an ideal place to camp and we suggested that they headed for it without further delay.

Roy rode on ahead and we followed in low gear, keeping a watchful eye on the tandem and trailer as they wove their meandering, bouncing way at the side of the road.

We waited to see the boys pitch their tent.

"We've got a bit of food to finish up," said Roy, indicating the unpacked tins of pilchards, corn beef, raspberries and cream, which it seemed, they intended to devour in one mammoth meal.

"Now be on the road first thing in the morning," was Dug's final instruction.

Sure enough I had just finished dressing when I heard the squeak and grind of the tandem at the top of the hill and when I got to the kitchen Roy's head appeared round the door.

"We're back and all's well," he announced.

Much to our relief, the tandem, after that one excursion was never seen again. Perhaps it was returned to the scrap heap.

The following Saturday morning I was making pastry when Nick burst in at the back door.

"Come and see what we've got now" he said.

I hesitated, looking at my floured hands.

"You must come," he insisted.

In the yard I found Roy, Tony and Peter standing in an old-fashioned high wheeled milk float, newly painted black and yellow. Tony held the reins of a mettlesome bay pony in the shafts.

"He's Flash," Peter told me.

I thought the whole turnout was pretty flash.

"He's Tony's pony and his father bought the float from the gypsies."

I duly admired both, stroking Flash and swotting at the flies, which seemed especially numerous that morning. They seemed to be collecting not so much round the pony as round Roy and I rather suspected that the latest addition to his hat, a fine mole skin, was not properly cured.

"Like to drive it up the road?" Tony asked me.

I had driven a milk float like this back in 1941, with a six-teen-gallon churn with brass plunger up in the front and I had not driven a horse since the war. It would be pleasant to take the reins again behind this lively bit of horseflesh... but thought of my half made pastry and cakes almost ready to come out of the oven.

"Another time," I said.

During the weeks that followed I often met this turnout along the lanes, the front filled to overflowing with Roy and Tony and Peter and Keith and Tom, until I feared that Flash might share the fate of Tom Pierce's grey mare, but he seemed well able to cope with his load.

But in time this too palled. Flash was permanently turned out in his field and the float remained in Tony's father's garage where it may still be today. Like the tandem, I never saw it again.

15. *To Put Away Childish Things*

As in most families there came a time when everything seemed to happen at once. For years we had grown accustomed to the routine of school terms and holidays, the familiar faces of the gang and to their unpredictable activities. Then, in the space of a year or two, everything changed. By the spring of 1964 Nigel, with an agricultural career in mind, was doing his first year's practical on a nearby farm. His last year at school, as head boy, had given him a good sense of responsibility but, academically, with less than the number of 'O' levels and possible 'A' levels required by most Agricultural Colleges. Painstakingly I wrote to them all, even to Sudley, receiving a reply from the latter, that they would be delighted to receive my son were they not entirely a college for young ladies!

Oh well, we all make mistakes.

Finally, he got a place at Sparsholt Farm Institute (now Sparsholt Agricultural College) with a provisional place at Shuttleworth, pending the results of his exams.

Meanwhile Nick had left school and not having decided on a career was taking a Business Studies OND Course at Chichester College of Further Education, working on the petrol pumps at the local garage in his spare time. One result of this change from school to college and jobs was that the humble pushbike was no longer considered an adequate means of transport and we were into the scooter era, with its attendant problems.

Funds being low, Nick picked up cheaply a somewhat decrepit machine, which required a steady supply of spare parts to keep it going. Purchase of these consumed much of his part-time earnings, fitting them most of his spare time, while fetching them from a garage which specialised in these things, which was several miles away, entailed many trips in the car or Land Rover. It seemed, at first, that Nigel had done rather better when he turned up with a large bright red American machine in apparently good condition, but it soon became evident that the engine was too high powered for our narrow, winding lanes.

It was named an Albatross, rather aptly as it turned out, for if the Ancient Mariner had trouble with his bird of that name it was nothing to the chapter of accidents Nigel experienced with the scooter. I constantly administered first aid to grazed knees and lacerated elbows.

One evening Roy arrived in the yard, not on a second hand scooter, but a brand new Honda motorbike.

"Come for a burn up on your Albatross?" he asked.

Nigel agreed.

Half-an-hour later Roy returned alone. Nigel had gone off the road and hit a tree. His helmet had saved him from serious damage but he had a cut face and was badly shaken. Would I come and pick him up? I hastily got out the car and drove with Roy a few miles where, by a sharp S bend Nigel was sitting by the tree, bleeding profusely from a gash in his left cheek. The scooter was undamaged, so leaving Roy to ride it home, I took Nigel in to hospital to have the wound stitched. It healed well, but I noticed, in the weeks that followed, how he frequently examined the cheek long and anxiously in the mirror, until I was moved to remark that when my own face was a mass of similar wounds I had not shown so much concern.

"Huh, it was alright for you," replied Peter defensively, overhearing my remark, "You were already married."

Apparently, once you have got your man, or your woman, it doesn't matter a damn what happens to your face.

That episode was not the end of our troubles with the Albatross. A few weeks later we had just sat down to our first feed of strawberries with some cream that had been sent us from Devonshire. There was time to enjoy them leisurely before setting off to a Confirmation Service at the village church, where Tony was to be a candidate. Dug, as Rector's Warden and I, had been invited to a small party at the Rectory beforehand to meet the Bishop.

"Has Tony gone home?" I asked, knowing that he had been with the boys all the afternoon.

"Not yet," replied Nigel. "He's just asked me if he could have a ride up the road on the Albatross."

Dug and I barely had time to exchange glances of apprehension when there was a tap at the door. When I opened it Tony was leaning, limply white against the jamb, blood oozing through the ripped leg of his jeans.

"I came off," was his faint explanation.

I sat him down and gave him some hot sweet tea and an aspirin before starting work on the leg, which was one deep abrasion from thigh to ankle, first to wash out the muck and grit from the farm road. Luckily he had had a tetanus injection just before he left Kenya. After I'd dressed the leg with a sort of patchwork quilt of penicillin gauze, Dug took him home. I glanced at the clock. There was just fifteen minutes before we were due at the Rectory.

"I can't make it," I said to Dug as he went out of the door.

"All right," he replied. "I'll convey your apologies."

I looked round for my strawberries and cream, not that I had much appetite for them. They had gone. The boys, anticipating my loss of appetite, had decided to help me out.

After I had gulped down a cup of lukewarm tea, I changed my mind. I would just have time to change before Dug returned and shortly afterwards we arrived, calm and smiling, to meet the Bishop.

Tony turned up at the service. I sat next to his mother and we both anxiously watched as he limped painfully, with ashen face, up to the altar, but he stuck it out to the end.

A few weeks later when Nigel sold the Albatross, many shared the relief.

When cars began to replace scooters, it was inevitable that the role of driving instructor should fall to me. I was considered the most fitting, having passed a driving test a few years previously, thus remembering fairly accurately what was required, and because my nerves were either of cast iron or had been stoned dead years before.

The stretch of road between Court Hill and North Wood buildings was a favourite place for all first lessons. The only hazard was meeting one of the farmer's sons from Gumber who were fast and furious drivers.

"If you see the Gumber van coming," I warned Nigel, "pull into the side and stop."

Our first two trips were uneventful. The third time we were halfway along the lane when the familiar green van came hurtling toward us.

Nigel remembered the 'pull in' bit but not the 'stop'. As he dived to the side of the road without decelerating, we hit a series of deep drains that run out from the wood, my head hitting the roof at each bounce until he at last responded to my agonised cries to 'STOP'.

Somehow driving lessons had to be fitted in with my already overloaded routine.

"What about a driving practice, Mum?" Nigel would enquire brightly each evening as I was washing up the tea things with my mind on getting Jonathan bathed and to bed before tackling that ever mounting pile of ironing.

"OK," I would reply, going off to find a volunteer to cope with Jon. Nick or Peter were usually willing but when I returned home I would rarely find that their efforts had been fruitful. Jon was usually running round the lawn in his pyjamas, rather dirtier than before his bath and in a state of excitement that eliminated the possibility of sleep. It seemed that this sort of thing went on for years, for no sooner had Nigel passed his test than it was Nick at the wheel. He was an entirely different pupil. Whereas Nigel had tended to drive at a tension, Nick was so relaxed I feared that he might fall asleep and I had difficulty in persuading him that an examiner would expect him to keep both hands on the wheel. I knew, the following year, that I would have few problems with Peter, for he had been a back seat driver for years. It was only a matter of sitting in the passenger seat while he practiced. It was with infinite relief that I dropped my last learner at the test centre, uttering those few final well chosen words of advice and caution before setting off on a shopping expedition, to which I gave only half my mind and to be met on my return with a triumphant grin and a car minus the L-plates, which had seemed a permanent part of its equipment.

But to go back to the September of '65, Nigel, after a few lessons from a qualified instructor added to my efforts, was able to take the wheel and went off in our old Hillman at the beginning of his first term at Sparsholt.

Our first visit to see him settled in was a nostalgic one, for the Institute's fields adjoined those of the 5,000 acre farm where Dug and I had worked and met back in '44.

Nothing had changed. There was the farmhouse on the hill were I billeted. The pub by the cross roads where Dug and I waited for the Blue Empress to take us into Winchester, a stout-hearted little bus that, crammed with passengers to the step, groaned and steamed and panted up the steep incline.

By those crossroads I waited on my tractor before D-Day while long convoys of army lorries packed with troops trundled past on their way to the coast, while planes and gliders flew overhead.

We went into the little 12th century church where we were married and looked at the cottage where we had started our life together. Even in Winchester everything looked much the same. When Nigel met an attractive schoolteacher, they frequented the same haunts, the cinema, the same place to eat, the long walk through riverside meadows to St Cross that we had known more than two decades before.

That summer virtually saw the end of the gang. The club, after two successful seasons had disbanded, the pop group first, quite suddenly, after the lead guitarist had absconded with the funds.

The boys that had made this house a second home since they were in short trousers were boys no longer, but young men intent on the serious business of pending careers or even already earning their living. Tom had a job in a bank, Keith was apprenticed to an electrician. That autumn Roy was off to Aberystwyth University and Tony to Edinburgh.

When they met, they were serious and adult, discussing jobs, careers and examinations. Sadly, it seemed that for the time at least, they had closed the door on the frivolities of youth as surely as the door was closed upon the cellar from whence their laughter and gay banter had issued with the constant flow of pop music. The cellar was abandoned once more to the damp, the dust and the spiders. The posters and the pictures of pop singers had damply peeled and fallen. Only a mildewed Mick Jagger clung tenaciously to the wall.

All this was very disappointing for Jonathan. Accustomed to so much amusement and attention from the older boys, he now found himself left very much to his own devices. Even Peter, the only one still at school was occupied with swotting for exams.

This left one busy housewife and mother at her wits end to entertain one bored, disgruntled boy of four.

A friend who told me of a small pre-preparatory school about four miles away saved me. It was run with great efficiency by a thin, bespectacled lady of rather severe appearance and middle years. Her school, she said, was not to be confused with play or nursery school. They got down to serious hard work. I looked down at the wide-eyed, round cheeked faces of the three-year-olds and thought of all the entrance exams, 'O' levels and 'A' levels, perhaps university degrees ahead of them and felt that serious hard work at such a tender age was a little tough, but I could see that she meant what she said.

To paraphrase the Duke of Wellington, "I don't know what she did to the children, but my God she terrified me." She was, all the same, a good woman and an excellent teacher. Jon learned quite fast and there was a distinct improvement in his general behaviour. As several children from the surrounding area attended the school, I joined a rota with three other mothers to transport six three to nine year olds to and fro.

Every morning, on alternate weeks, Jon was picked up in a large Rover already full of lively children, including the driver's youngest, a baby of nine months in his special car seat. I appreciated the amount of effort that must have gone into getting her own children ready and collecting the others at quite an early hour and I admired that young mother's sweet and smiling composure that was as unfailing as the morning. On alternate weeks it was the turn of our vet's wife, pretty, slim and businesslike. Mine was an afternoon run, the children, released from the restrictions of school and recognising in me one not very able at applying discipline, released their high spirits in riotous games of cowboys and Indians over and between the seats.

Unfortunately, not long after Jonathan joined the school it closed down, although to my knowledge there was no connection between these two events. However, he was now almost five years old and soon able to go to the village school where his brothers had begun their education.

Somewhere about this time all the sheep were sold off the farm. We would miss the familiar scene of the flock grazing on the Downs or being driven past the house to the sheep dip by the dairy, spring lambs gambolling in the field and old Shep gently driving his cart along the lane.

Early on the second Saturday in September they were carefully graded and loaded into lorries, keeping the grades separated, for Findon Sheep Fair.

Shep and Ethel rode in one of the lorries, Dug and I with Jonathan following by car. It was a typical September day, starting with an autumnal nip in the air and a slight mist over the Downs soon to be dispersed by a sun, which still held the warmth of summer.

Turning off the Worthing road a few miles past Arundel we were soon in rolling chalk downland where well-farmed acres of stubble and plough were spread before us, unbroken by hedges and only sparsely wooded with small copses. Reaching Findon itself, centred in a hollow, we climbed out again through its steep narrow streets, flanked by picturesque cottages and gardens, into a wooded, flint walled lane, which came out at Nepcote Green. Here sheep fairs had been held since 1790, not only South Downs but breeds from all over the South of England being represented and as many as 20,000 sheep changing hands. The lorries carrying our sheep pulled alongside the many others parked near the long lines of sheep pens. Dug drew up on the verge and we got out and looked about us. On the green stretched a solid expanse of woolly backs broken only by the hurdles and the gangway where farmers and their shepherds stood with their dogs, discussing the days' business.

To the East rose the great prehistoric hill fort, Cissbury Ring. To the North lay the training gallops of Findon Training Stables.

Having seen the sheep penned, again in their separate grades, we stood for a while talking to Shep and Ethel, who introduced us to their brother George, an older and more weatherbeaten version of Shep. A shepherd on the Downs for fifty years, he remembered the sheep fairs before the days of the lorries. Then the shepherds walked their flocks as far as twenty miles across the Downs. It was a pretty sight, he said, to see the sheep converging on the village from all directions. After the sale, the shepherds had a jolly reunion in the Gun Inn. George had a rare collection of sheep bells, which he invited us to go and see when we were next in the vicinity of his cottage.

At the further end of the green was the fairground and seeking some amusement for Jonathan, I made my way there. It was a small fair arranged in a neat circle, but it contained all the essentials, a roundabout with handsome prancing horses, a coconut shy, shooting range, toffee apples and candy floss.

We stood watching a gang of small boys at the hoop-la. They spent their money and moved on.

"Come on, have a go, Gran," cried the young man at the stall.

I looked round for Gran. Beside me was a teenage girl with a small child and behind me a young man; no one else. He must have meant me. I looked at him coldly and walked away. As I did so I caught sight of myself in a mirror on the shooting range. Did I look old enough to be Jon's grandmother? My hair, except for silver bits at the side was still dark and my skin, with its summer tan looked reasonably young. Come December, my winter face might be a different story, but I still weighed no more than eight and a half stone. Nevertheless I was old enough. I might have had a daughter when I was twenty who had a child when she was the same age.

I was spared further contemplation on the subject by a sudden commotion behind me. To the ceaseless, many toned bleating of the penned sheep was added a nearer, more urgent note, together with the drumming of cloven hooves. About twenty or thirty sheep were rushing toward the fair ground like the herd of Galilean swine. Women snatched up their children, fair men rushed to defend their stalls, but the sheep ran between them and behind them.

Stands rocked and prizes toppled; never had so many coconuts fallen in so short a time. About ten ewes dived under the roundabout and for a while no amount of shouting and prodding would evict them.

I had fancied there was something familiar about these errant sheep and the sight of Shep and Ethel running towards them, waving their arms like two demented scarecrows, confirmed my fears. They were ours. Then Dug came up with two more helpers and gradually the runaways were rounded up but no longer graded and the ones from under the roundabout had backs liberally daubed with thick black grease from the machinery. I never quite knew how they escaped from their pens or what happened afterwards. Perhaps they were sold as a mixed lot. As we drove home with Shep and Ethel in the back of the car I did not think it tactful to ask too many questions on the subject.

Happily, the selling of the flock did not mean a complete farewell to Shep and his sister. Dug found them a job on a neighbouring estate with a flock of pedigree South Downs and often when driving home from Chichester I would meet Shep pushing his bicycle and stop to chat. I would enquire after Ethel's arthritis and the welfare of the flock and he would proudly produce from his waistcoat pocket some much fingered snapshots of himself with a rosette-bedecked ewe at the Royal Show.

We never did see George's sheep bells. When we visited him he had sold them to the landlord of a pub in Amberley. Perhaps we would look at them there some day.

16. A Scandinavian Interlude

One morning in February the telephone rang just as I was dishing up the lunch. Expecting that, as usual, it was someone trying to catch Dug as he came in to a meal, I was surprised to hear Nigel's voice. There was something in the tone of his "Hello Mum" that made me apprehensive. Sure enough, "I've had a slight mishap," was his next remark.

Knowing from experience that 'a slight mishap' could be anything from getting a girlfriend locked out to wrecking the family car I just said "Oh" and waited. It was a car crash, but as it turned out, not too serious and not his fault. He had been waiting to pull out at a junction, when a van, in trying to avoid a clergyman doing a three-point turn in his saloon in the middle of the duel carriageway, had crashed into him. Neither he or his girl companion were hurt, nor the van driver although the van had finished up halfway through a hedge, but the left wing of his car was completely smashed.

I just had time to say how sorry I was at his bad luck, but at the same time relieved that no one was hurt and that I would have expected more consideration from a gentleman of the cloth, when Dug came in and took over the phone.

As it was quite clear which party was in the wrong, there was no difficulty about the insurance, but all the same it was a slow business and Nigel was without a car for many weeks.

Accidents, like measles, seemed to run through the family. It was my turn next. Now that Jonathan was at school, I was having a second round of attending parties, birthday and un-birthday, for the young. It was on one of these occasions that I had a 'slight mishap'.

Having deposited Jonathan at the party I hurried away to pick up another young guest and mother who had no transport of her own. It was a teeming wet day and carefully skirting a spreading lake in the middle of the drive and averting my face from the driving rain, I jumped into the car, slapped her into reverse and accelerated. There was a sickening crash and through my misted window I saw something grey leap up onto my right wing.

I scrambled out of the passenger door to see a small van perched on the mudguard of the Hillman. At the same time two men approached from a shed in the garden. A tall, hefty, dark-haired young man and his small rather wizened companion. Even in that fraught moment their resemblance to Harold Steptoe and his aged father did not escape me.

"Is it your van?" I asked them. "I am terribly sorry."

They did not reply, but stood in silent contemplation of the two vehicles while the rain continued to lash down upon the three of us. Then the young man, flexing his big shoulders lifted the van off my car and let it drop. It sat there trembling in the wet, while Harold slowly stroked his chin. Then he took hold of the crumbled mudguards, lifted them and let them go. They flapped dismally. The little grey vehicle looked like Disney's Dumbo about to take off.

"There ain't much you can do to that, that hasn't already bin done," he remarked laconically.

The older man moved forward then, pushed back his cloth cap and scatched his balding head. He looked at Harold and the two shrugged their shoulders.

"Well my dear," he said kindly. "It's a grey van and it's a grey day, I reckon you just didn't see it."

I reckoned that is how it was too, but it was jolly nice of them to take it so well. I did not meet such magnanimity when Dug saw the streaks of grey paint on the new Hillman.

A few days later Nick, on his scooter, mounted the boot of a car going over Arundel Bridge. The driver was very put out although no damage was done to his vehicle but the scooter was put out of action for a week. During this time, Nick decided as he now had a job in the accounts department of a local firm of hauliers, that he could afford an 'old banger'. It was a 1948 black Morris Minor and when I first saw it, Nick was driving it round the front lawn.

Due to some peculiarity of the steering it tended to proceed in a rumba-like side-to-side motion, which increased with its speed and with the number of its passengers. When these included all the remaining members of the gang, seated upon each other's knees, its progress was extraordinary to behold. But it had a splendid engine and went 'like a bomb'. It carried Nick to and from work, to the beach and occasionally to London without a breakdown and did an incredible mileage during the summer months.

Then one evening I listened for the familiar roar and rattle of the old banger bringing Nick home to tea. Instead there came something heavy and smooth and so subdued in tone that I scarcely heard it. I looked out to see Nick, an expression of barely suppressed pride and elation on his face, alighting from a black, shining Riley 1½ litre.

"Where did you get that?" I asked.

"It's been in the local garage for weeks. I've looked at it every day. Today I couldn't resist it and went in and made an offer for it. I had almost enough money saved up and the salesman said I could have it £10 cheaper if I didn't ask him to take in the old banger."

I could see his point.

Together Nick and I admired the engine, the excellent condition of the coachwork, the real leather upholstery and the grained wood of the doorframes and facia; nothing synthetic there.

"Like a run up the road?"

In contrast to his hurtling along in the 'banger', Nick drove his new possession with care and dignity. Such comfort and stately progress made me feel like a dowager duchess.

The petrol tank held twelve gallons but the most Nick's finances would run to at a time that summer, was one. The girl on the petrol pump at the garage, which received his regular if not extensive custom, soon remembered him. "Another whole gallon?" she would enquire sweetly with pump hose poised. Some weeks later Nick surprised us with another purchase.

"I've bought another Riley," he said casually one teatime, very much as he might have told us, some years before that he had bought another rabbit.

"Whatever for? What are you going to do with it?" asked Dug.

"Well, it's got a splendid engine. I thought I'd make a hot rod of it to drive over the Downs. It was only £10, an absolute cinch, a 2½ litre." I could only suppose that a recent birthday accounted for the improved state of his pocket.

"It isn't licensed or insured so could you tow it over from Bognor for me one night Dad?"

Dug agreed with some reluctance and the next evening the Land Rover pulled into the yard towing a large grey car. It was older than the first Riley but in good condition and the engine was perfect. A trial run up the farm lane proved that it would do 75mph in third gear, fairly purring along.

"That's a damn good car," Dug exclaimed. "You're surely not going to pull that to pieces?"

Nick said nothing, which meant that was exactly what he intended to do and the following Saturday having got together a few helpmates and a tractor he got to work. At the sound of tearing metal I looked out to see the coachwork being ripped off the chassis. It was a sight that would have reduced a Riley enthusiast to tears. All that remained at the end of the morning's work was the chassis, the engine and one seat. It was then towed away to a small patch of grass behind the building to await the next time Nick had the opportunity to work on it.

Somehow that opportunity never came. The denuded car stood there, the engine barely protected from the elements by the lightly replaced bonnet. In the spring, a blackbird built her nest under the clutch pedal and returned there year after year. Nick's plans for a hot rod never reached fruition and in the end Dug sold the Riley to the scrap man for £5.

While Nick was tinkering with cars, Nigel was having more serious thoughts about his future. His year at Sparsholt was at an end. In July we attended the college open day. As we walked round looking at the livestock and the hothouses he told us that although he felt that he had done satisfactorily in his exams (he had in fact passed all that he had taken), he had decided to give up his place at Shuttleworth and take a job. Several of his fellow students were going to farms in Denmark and he had applied to the Danish Sponsoring Authority.

A few days later he telephoned to say that he had got a job as an assistant on a 150-acre farm at Skjod in Jutland. He must leave in two weeks time and would I book him a passage, anything would do.

Much of the remainder of that day I spent on the telephone to all the travel agencies in the area. They all said the same thing. There were no passages to be had at such short notice but if I cared to keep phoning I might be lucky in getting a cancellation.

On the third day of trying the Chichester Agent came up with a first class double cabin. Well, Nigel had said anything.

He came home on the third Saturday in July and was to sail the following Thursday. It was natural that during those few days life should revolve round the preparations for his departure: passport and photos to be got in haste, clothes to buy and a long white canvas bag like a sailors kit-bag into which, with great determination he crammed all the possessions he intended to take with him. It was re-crammed several times with varying arrangements to make better use of the limited space and a couple of times more because there was inevitably something packed that he needed. Thursday came, and with only half-an-hour to the train departure time and Dug at the door with the car there was one last frantic unpack because at the bottom of the bag was the tie he wanted to wear.

Even so, we all three wore a resolute calm, even Dug, who liked to allow plenty of time for catching trains and fretted excessively at delays. As Nigel threw the bag into the back of the car we could hear his alarm clock ticking from its pregnant depths (I hoped it would not give rise to suspicion on the journey) but his wrist watch he had left in the bathroom.

"Here, take mine," said Dug, determined to brook no further setbacks. The train was at the station when we arrived. There was just time for Nigel to get a ticket and grab a seat before it moved out.

I followed him mentally across London and up to Harwich as I went about the chores that day and that night when an unseasonable gale battered the windows and tore the newly formed apples off the trees, I thought of him on the North Sea.

But I needn't have worried. In his first letter home there was no mention of a storm, only of a charming Danish girl whose company he had enjoyed during the crossing.

At home Nick, having obtained a Business Studies OND had decided on a career in computers. After several applications and interviews he landed a job as trainee operator at the Imperial College of Science and Technology.

Since Roy and Tony had left for their respective Universities, Nick's chief companion had been Mike, who, like the others, had made the farm their second home and had been living with us during the harvest while helping on the farm. He was in printing and having finished his training was, like Nick, about to take a job in London. It seemed the obvious thing that they should take a flat together. They found one in Sinclair Road, furnished but not fully equipped. Having seen Nigel away I set about turning out cupboards and drawers to see what bed linen, cutlery, pots and pans could be spared for the flat. One Sunday Nick, with a loaded Riley, set off for London.

Number two had left home; now we were four.

Nigel's letters came regularly from Denmark. He found the life quiet but enjoyable and inexpensive. The working day began at 7.30 and ended at 5.30. He seemed happier than the other seven students who had gone out to Denmark from Sparsholt, none of whom stayed more than a few months. He perhaps had the advantage of working on a small farm, experiencing a variety of jobs and living with the farmer and his wife, who both spoke English. The others on large farms complained of the tedium of spending many weeks at a time on one job, of milking in large herds at 3.30am and the loneliness of working with other students who did not speak the same language.

In November the first snow fell in Jutland. By this time Nigel had bought himself a scooter and was attending night classes in Arhus to learn Danish. Here he met an attractive Danish girl, called Inge. It seemed that life was pleasant enough.

Then one week there was no letter. They had arrived regularly every week up to then but there might be several reasons for the delay. His last letter had reported gales following the snow that had blown down ninety-six telegraph poles across Jutland, more heavy snow and long hours milking by hand when the electricity failed was enough cause to put off writing.

The following week a letter came which told us that he had run out of road when riding his scooter back from Arhus and as a result had spent three days in Arhus Hospital. It was not until some weeks later that we heard all the details and realized the gravity of the accident.

Having discovered on a map a shorter route to Inge's home he decided to try it out. On the way he came upon a sharp S bend which he only just managed to negotiate. On the return he forgot this dangerous piece of road and ended up in the hedge, where he lay for two hours unconscious with a broken hand and bleeding profusely from a gash in his head, until discovered by two young Danes.

Only several pints of Danish blood and the excellent hospital care for which he was full of praise, saved his life. Later he sent us a cutting from the local newspaper which read (translated) 'English student found without life by roadside'.

It was many weeks before he had full use of the hand, which left Finn Hansen without help in the cowshed. This fact seemed to be Nigel's chief concern and further letters reflected his depression until he was able bodied again.

At Christmas Peter left school and started work on the farm although not with an agricultural career in mind.

Since his earliest years, when he had constantly driven dinky cars along floors, over sofas and tables making the appropriate noises, his greatest interest had been the motorcar. As soon as he had saved enough to buy his first car and passed his driving test he intended to train as a car salesman.

The remainder of the winter passed uneventfully. Jonathan had settled down at the village school. I met him each day as I had the others ten years and more ago but waiting with a different group of Mums and with a different Golden Retriever. In the evenings, when the days got longer, I started my third stint as a driving instructor.

Nigel wrote to say that he was changing his job in the spring. Although his relationship with Finn Hanson had continued to be of the best, he wanted to see another part of Denmark and to get more experience.

As he intended to remain in Denmark for some time to come, we decided to go out and see him for his 21st birthday, which was in June. Nearer the time he found us hotel accommodation at Naestved, a small town a few miles from the 200-acre farm where he now worked for Hans Jorgensen. On 31st May we sailed from Harwich in the *England*, one of Denmark's beautifully equipped and spotlessly clean ships.

Our first glimpse of Danish scenery as we travelled by train from Esberg across Jutland was unexciting. It's green fields full of Friesian cattle resembled some parts of England, except for it's hedges of lilac bushes, but gradually in the days that followed we fell under the spell of Scandinavia, it's pure air and clear brilliant light and scenery that became increasingly varied as we discovered it.

The weather was as changeable as it is at home. On the evening of our arrival as Nigel showed us round Naestved we shivered in an icy wind.

"They say it blows straight from Siberia," he said. We believed him. But the next day when we went to collect the car Nigel had hired for us, the temperature was in the 70s. We found our way to a little resort on the Baltic with the unpronounceable name of Karrebaekminde and roasted ourselves upon the sand. The Danes, even small children, were bathing happily and I wished I had brought the swimsuit I had hopefully included in my luggage.

"Oh, the Baltic is too cold for an Englishwoman," said the hotel proprietor that evening. Having put my feet in I had not considered it any colder than our Atlantic coast in May or the Channel in October and I was determined the next day to prove him wrong.

But the following day when I was armed with swimming tackle we took a wrong turn and came out at a different part of the coast. Here there were trees and sand dunes and rushes edging the water making it look more like the shores of a lake.

There was not a soul in sight and a large notice said '(something) forbundt' The first word was partially obliterated leaving us in doubt as to what was forbidden. Perhaps it was bathing, so I did not swim. Later we learned that it was parking and it was by this sign that we left the car.

Although the next two days continued to be hot, there were other places to visit beside beaches. Very near Naestved was Gavn, the Castle on the Island, with its ornate chapel, art collection and glorious flower gardens, then ablaze with tulips; Gisselfeld, once a nunnery, set against a mystic lake and acres of arboretum; The Island of Mon, famous for it's spectacular white cliffs, the only ones in Denmark.

The day after the weather changed. It was cold and wet.

I never did get my swim in the Baltic.

On the Sunday when Nigel had the day off and took us to Copenhagen, there was a chilling mist and a fine typically English drizzle, perhaps appropriate conditions for visiting eerie Elsinore.

During our stay we were entertained by Mr and Mrs Jorgensen. On our arrival, after we had been welcomed cordially by our host and hostess, their five children appeared, quite suddenly from divers directions and lined up punctiliously to greet us with the bobs and bows and handshakes, reminiscent of the Von Trapp children in 'The Sound of Music'. We did not see them again during the evening for they did not dine with us but had their meal in the kitchen from whatever came out on the dishes from our table. Reflecting how careful we are at home to get our children and our dogs fed before the arrival of guests I wondered if we pampered them too much.

I was impressed by the independence of Denmark's children. Even the smallest cycled a considerable distance to and from school.

"We do not run about the whole time after our children like you English mothers," remarked Mrs Jorgensen, who was tall, blonde, vivacious and very energetic and had been an au pair in England before her marriage. After school they took their hoes and went out to the field to work until the evening meal. They were industrious, extremely well mannered and happy.

Mrs Jorgensen, besides coping with her home and family, helping on the farm, serving on the board of governors of the local school, she bred Welsh Mountain Ponies. A member of the Danish Welsh Pony Society, which has done much to improve the breed in Denmark, she had a stallion and three mares imported from North Wales, two of which, had that year both produced filly foals.

Finding us keenly interested she took us to see her stock and we spent the rest of the evening looking at photographs and various data concerned with activities of the society.

It was with regret that we left Denmark after ten days. Mr and Mrs Jorgensen and their entire family came to the station with Nigel to see us off at the start of our return journey.

Dug and I were quite possibly the world's worst travellers. If it is possible to take the wrong plane, boat or train or even get on the wrong local bus, we would do so without a moment's hesitation. Even on a trip from Portsmouth to the Isle of Wight we contrived to do the crossing on separate boats at an hours interval.

The fact that we arrived in Naestved without a hitch was due to a charming Danish sea Captain who travelled with us to Korsor and steered us through our various changes of boat and train and after he left us, to a farmer with whom we got into conversation and who was meeting his English au pair. He did not find her, but he offered to drive us to his home-town, which was on our route and put us on the train there to Naestved.

Travelling in his fast and comfortable car was relaxing and he was not unrewarded for his kindness, for when he took us onto the station platform, there waiting was his au pair.

On the way back we were on our own and it was at Korsor that we first went astray. Although according to our timetable we had two hours to wait, when a boat came in and other waiting passengers boarded her, like two sheep we followed them. It was not long before we realized that this was a far superior vessel to the train ferry we should have been on. We went below to look for the train. As we thought, there wasn't one. On deck again we saw the name; the *Princess Beatrix*. No one, however, contested our right to be on her so we relaxed in her comfortable lounge and ate delicious *smorrerbrods* in her smart refreshment room.

When we docked at Fredercia instead of Nyborg we got on a train, which we eventually discovered was not going across to Esberg but North to Arhus. Still no one had checked our tickets and we saw a whole lot of Jutland at the railways expense. After a hasty and expensive cup of tea on Arhus station we at last took a train to Esberg, arriving there at precisely the same time as the boat train we should have waited two hours for at Korsor and in good time we embarked for our return voyage in the *England*.

Before we left Denmark, Nigel had told us that he was leaving the Jorgensens the following week and going to Norway. A young Yorkshireman by the name of Andrew whom he had met in Arhus was at the International Hostel at Valdall and working in the area and Nigel was going to join him there.

During his last week at Sibberup, he had a welcome change from the tedium of milking and hoeing to forestry work, sawing and loading clipped trees and taking them to the National sawmills. Some timber he took back to the farm for building. He assisted too in the planting of two hundred and fifty fir trees alongside the pony paddock for winter protection.

His first letter from Norway arrived during the third week of June.

"Well, here I am after thirty-four hours of travelling. Valdall is a self-contained village separated from Aandalnes, the nearest town, by fifty miles of high mountains. The bus from Andalnes zig-zagged up a small one with thirty-hairpin bends. The road had only been opened two weeks. It is closed by snow for eight months of the year when the alternative is by boat".

The hostel was a modern building and to date there were seven other young men staying there. Two from Australians from the backwoods, who had hitch hiked from Ceylon, one Canadian, three Scots and Andrew. They were not allowed to stay, however, unless they had work and this presented a problem, as the Norwegian farmers were reluctant to employ them. Only Andy, whom they had found to be a hard and honest worker, was inundated with offers of employment.

"He had tried to pass some of the work onto me," wrote Nigel, *"but the farmers will not take me on. If only I can get a chance to prove myself on one job, I'll be in."*

At last Andy forced his employer's hand by threatening to leave if Nigel was not taken on too. A few weeks later when Andy went to work on an island below Alsund for a while, unloading fish boats for an oil and cake factory at £30 a week, Nigel took over his job of haymaking at 5/- an hour with lunch and a bicycle supplied.

Meanwhile he did a variety of jobs. Painting barns and chalets, taking cows up the mountain (four hours the return trip) and cutting grass round the cherry trees on the steep slopes with a large hand machine. Most jobs were at least a half hours walk from the hostel and required taking a packed lunch.

203

His only complaints at that time was the rain that fell for three parts of the time the humidity, which was hard to work in and the twenty-four hours of near-perfect daylight.

When strawberry time arrived (that valley grew the best in Norway) Nigel and Andy were employed in picking them and cherries for 5/6 an hour or 6/- a pound. They helped unload twenty-five tons of Polish sugar, only a fraction of what was used in the season in the jam factories. (On his return home Nigel brought some of the Norwegian strawberry jam, the most delicious we had ever tasted).

"Life is still very quiet," he wrote at the end of July.

> *"A few chaps go fishing at night but all the salmon caught in the rivers is sold. Usually there is nothing to do but sit and talk but this is pretty entertaining. I've acquired valuable information about the USA, Europe and West Asia from students who have hiked through these countries. Many of them just pass through, staying one night. Last night there were two girls who had hiked from New Zealand. Interesting listening to the tales of their travels. Last Saturday there was a dance in the village at the top of the fiord. We all went in rowing boats or the doctor's motor boat".*

The following week there was a party among a group of *soates* (weekend cabins) After a two-hour climb they arrived at the place to find eighty youths and girls singing to guitars and drinking homemade wine and beer round a huge bonfire.

In August we learned of his plans to return home at the end of the month. He and Andy were buying a van and travelling through Scandinavia and Europe. From acquaintances they had made with several girls who had stopped at the hostel they had invitations to stay in Oslo, Holland and Belgium. They had yet to find a friend in Germany. Nigel was not sure how long the trip would take them but we could expect him, with Andy, sometime in mid-September.

17. The Second Time Around

We arrived home in June, just in time for Peter's driving test. I left him at the test centre to do a round of the shops, pretty confident that he would pass and, sure enough, when I returned he was taking the L-plates from the car.

For the third time we were in the position of sharing the current Hillman with a teenage son. This required a degree of tolerance and accommodation on both sides, but by the end of harvest Peter had saved enough money to buy his first car; not an 'old banger' but a very respectable Austin Countryman and the Hillman was ours again.

With only Peter and Jonathan at home full time, I now had a little leisure to take up writing again. Mrs Jorgensen had give me all the gen about the Danish Welsh Pony Society and a delightful photograph of three of her children leading three ponies at the previous years show at Odense. I had written out the rough draft of an article while sunbathing on the beach at Karresbaeksminde. At home I removed the dust from my typewriter and finished the feature. It appeared in September in *Horse and Hound*. Other ideas came crowding into my mind. What about the story of our cellar club? It was accepted by one of my old markets. I was back in business.

Meanwhile things continued to work out for Peter. When Dug took delivery of a new Land Rover it was brought to the farm by the sales manager of a firm in Brighton. Tall and elderly, with distinguished looking grey hair and beard he reminded me of the actor James Robertson Justice. Over a cup of coffee in the office, after business was concluded, he told Dug that he was retiring in a few years time and the firm was looking for a young man to train for his position. Dug said that his son was looking for a job with cars and agreeing that this might be just the right opening for him Mr J.R.J. waited until Peter came from the farm.

The following week he went for an interview. When he returned he not only had the job but also had found himself a one-room flatlet in Brighton. Two weeks later he moved in. With alarming suddenness another bird had flown.

After he had gone the house seemed strangely quiet and empty; he had always been the liveliest and the noisiest of the family. He was the only one to carry on an animated, if one-sided, conversation before a 6am breakfast, to which I responded in monosyllables in my somnolent daze by the stove.

That morning my story about the club had appeared in the *Farmer's Weekly*. All too vividly I had portrayed those lively years when the house had seethed with boys and girls and reverberated with their laughter and music. Jonathan, I reflected would miss out on companionship like that.

True there were still the weekends when both Nick and Peter came home. Nick was working at two jobs, nightshift on computers and as assistant accountant for another firm during the day. When home he studied for further examinations or caught up with some sleep. Sometimes Mike came with him and Tony or Keith or Tom would drift over. Then by Sunday night they were gone.

We were decimated in animals too. Jonathan's only pet was a hamster called Rusty (because of his colour), which it fell to me to clean out and feed. But he was rather endearing.

For the first time in many years we were without a dog, since Beauty had died in the spring and Jane some years before at the age of sixteen. We had sorely missed the old faithful, no other quite took her place and Beauty had pined for long while.

Now Ann had a four-month-old bitch puppy for sale and the next Sunday we went to see it. We liked Jessica and she settled down with us happily, immediately attaching herself to Jonathan. The two became inseparable.

Then there was Mog. She came in the manner of farm cats, as if from nowhere, following Dug to the dairy one morning when he went to fetch the milk and back again to the house. When he reached the kitchen door she stopped, looking up at him with mouth wide open in a pitiful, silent 'meow', then slipped past him. She was in. Having daintily accepted a saucer of milk, she sat down by the Rayburn and began to wash her already immaculate white feet and shirtfront. The rest of her coat was a pretty tortoiseshell, except for one ginger leg.

After that she spent all her days by the fire, watching, sleeping, washing, leaving only at dusk for nights of hunting in the farm buildings. Every morning, without fail, she met Dug at the door and walked with him to the dairy, mewing softly, talking to him, he said. He called her, simply Mog.

I liked to see her by the fire. She was so small and neat, exuding contentment. When I washed the floor I went carefully round her with the mop sponge, leaving her place until she had gone out. I liked to hear her vibrant purr, loud as an outboard motor. It was absurd from so small a creature. An American visitor laughed at the sound. "Jest listen to thet cat," she drawled, "She don't know she don't weigh a hundred and forty pounds."

A good mouser, she killed twenty-three one day and lined them up, not eating until we admired her prowess.

How hard she worked to feed her frequent litters of kittens. I once saw her drag a full-grown rabbit from the field, straddling it as a lioness does with her kill, for their hungry mouths.

Her relationship with Jessica was tenuous. Every morning they would greet each other fondly, Mog rubbing against the bitch's forelegs, her tail curling round the ruffed neck, purring sweetly while Jessica muzzled her gently. But sometimes, in a fit of jealousy perhaps, the bitch would seizes her by the scruff of the neck and shake her like a rat. Spitting and clawing, Mog would escape and flee to safety.

The next morning she would hide on top of the coal bunker in the well house waiting for Jess to come trotting in from the yard, then leap onto her back, claws extended. *Touché.* Then they were friends again.

Mog was with us for many more years. She grew old in her many litters of kittens, lost the last and became thin and limped from some old wound long forgotten. She seldom left the fireside. Then one day she seemed better, got up and followed Dug to the barn where she saw a rat in her path. With something of her old speed and expertise she seized and wrestled with it, biting the vital spot. But, as it lay dying, behind a sack of corn, so did she. There were other cats around the farm, gingers, greys, one tortoiseshell with a ginger leg, all descendants of Mog, but they did not come to the house, they were just farm cats.

I read somewhere that the animals one really cared for never die. Sometimes when I would pass the paddock, by some trick of the light, I could see Stobie half under the holly tree, the sun reflecting on the shining dapple of his flanks, or old Jane, sitting waiting by the stable door.

Heavens! Was I suffering from the dreaded 'empty nest' syndrome? I turned to the vegetables in the sink and looked out of the window as Jonathan went past. He was followed by boy after boy, I counted five. I looked outside. In the yard a pile of bicycles leaned against the wall. We were back to square one. I smiled. Who said one could never put back the clock. There was still the second time around…

The first time I met Daisy Hersee she was tapping the buttresses that jutted out from the wall of The Grange with her white stick. I had seen her often over the years, a small figure clad in black, on the arm of a friend; a sweet face framed by pretty white hair under the hat. I knew that, totally blind, she lived alone, cooking and performing all her household tasks with complete independence, but this was the first time I had seen her out alone and she seemed troubled by the buttresses. I asked if I could help and when I had walked her home she asked me in. That was the first of many visits; sometimes to read to her, often to just hear her talk.

Although partially blind since infancy, she had become companion and personal maid to Madame Bessie Belloc, mother of Hilaire, who had lived for many years in the village and had travelled with her each summer to her other home in La Celle St Cloud in northern France. Over eighty years of age, Miss Hersee recalled vividly a Slindon of over seventy years ago.

After the premature death of her mother she had been brought up by a Mrs Refoy, widow of a descendant of Samuel Refoy who built the folly above Court Hill Farm and Mrs Refoy's memories went back another fifty years, so that I was given a picture of Slindon back to the mid-19th century.

"There are not many of us now who remember these things," Miss Hersee said one day. "When we are gone they will be lost."

This, I thought, must not happen. Afterwards I always took a pen and note pad and wrote down everything she told me. That was the beginning of *Portrait of Slindon* – a local history of the village which was to occupy my leisure hours for the next four years and to go into five editions, each one longer that the previous one.[2] Many hours were spent in the archives and reference library in Chichester and in visiting other old people with past memories of the village.

Then I was lent the diaries of an estate worker, Jimmy Dean, who had died in 1935. These diaries, written in perfect copperplate learned at the village school, on identical Timothy White's notebooks, spanned fifty years and recorded events in Slindon about the farming, garden crops, weather and personalities. They were historical gems.

---

[2] The latest edition of *Portrait of Slindon* was published in 2002 by Woodfield Publishing – the first time it has been professionally published. It contains many new photographs and updated text bringing it up to date to the year 2002.

If you are writing a comprehensive local history, you must start with the geology and archaeology of the area, I was told.

How could I do this when these two subjects were a closed book to me? Then I thought of Martin Venables, well-known geologist and palaeontologist who was at that time curator of the Bognor Regis Museum. I had met him on occasion so now I went to see him at the museum and told him of my dilemma, confessing my complete ignorance of the two necessary subjects. Probably delighted to sow his knowledge on such virgin soil, he agreed to help me, as did his assistant, the archaeologist, Alan Outen, so between visits to Bognor's supermarket and meeting Jonathan from school, I exchanged my housewife's hat for that of a historian and learned (metaphorically) at their feet. In the school holidays Jon came with me and wandered happily round the museum looking at the exhibits. Mr Venables had been the first to stuff birds as if they were in flight and many of these hung from the ceiling and especially fascinated my young son.

There were others who came to learn, two young married students with a small baby and two boys with a little, shaggy dog, dripping with hair and seawater, who had found something of interest on the beach.

Another helper was Robin, tractor driver at Court Hill and an enthusiastic archaeologist who had amassed a vast collection of prehistoric artefacts and had an accurate knowledge of the field systems, Iron Age Forts and Roman remains in which the area abounded.

Eventually I completed my first chapter, 'Setting the scene'.

I often thought how much easier it must be for the male writer. He invariably had some devoted female with a typewriter, who would correct his mistakes, answer the telephone, make him cups of tea and generally smooth his path, whereas I did my own research, writing and typing with my eye on the time to get meals, collect Jon, feed the hens, walk the dogs not to mention answering the telephone, people at the door and taking messages to the farm. I had heard it said that a writer does his/her best work under pressure, so mine should be good.

I battled on.

18. And Girls

Into my hitherto completely male orientated life there now crept a little femininity in the form of the boy's girlfriends. When Nigel left Denmark for Norway he lost touch with Inge but after a few months there he wrote to say that he was coming home for a week and would be bringing Anna Mireta with him. In due course he arrived with a petite, dark-haired, pretty girl and her guitar. Each evening she entertained us charmingly with current popular songs accompanying herself on this instrument.

"Why don't we take Anna Mireta to the pub?" Peter suggested. "She could sing and we might get free drinks."

So she put on her national costume and entertained the locals in several pubs and they indeed got free drinks.

Sometime after her return to Norway, Nigel was invited to a return visit only, on arrival at her home, to be informed that she now had a Norwegian boyfriend so their relationship must end. Somewhat disgruntled he returned home and soon afterwards got a job as a farm assistant in Oxfordshire.

One day I answered the door to a large, rather plain girl who announced in a strong Scandinavian accent.

"I am Gunelang from Norway, I haf com to see Nigel."

Explaining as I asked her in, that Nigel was no longer at home, I gave her tea and we talked. She was staying with a party of young Norwegians at a hostel at Littlehampton. She had taken a bus to Arundel and walked from there. I showed her a little of the farm and took her to feed the hens and then drove her back to Arundel to get a bus back to Littlehampton.

"Oh, Gunelang," said Nigel when I told him. "Not one of the most attractive Norwegian girls, but a good friend."

Meanwhile Nick had given up his job doing computer work for the Greenwich County Council and gone into the wine trade; this involved trips to the continent and subsequently to girlfriends of various nationalities.

First there was Babe, from southern France who came for a week that first summer. One did not make the mistake twice of calling Babe French.

"I am not French, I am Basque," she would exclaim indignantly, making us aware of the fierce separatism in that part of France. Babe was critical of English cooking (mine in particular) and I considered she left much to be desired as a guest, but when, years later, she came to Nick's wedding with her husband and children I thought she had matured into a very nice woman.

Nick was never without a girlfriend and there were, in between continentals, a number of English girls. Louise, a loveable, lanky tomboy; Jeannie, a Scottish lass with a good brain (she became a barrister); two Barbaras at the same time (a confusing situation that required considerable tact on my part); Adele who cooked "film star eggs"; and for the longest time, Geraldine, a gorgeous international model.

At about this time, John, a friend of Nigel's since childhood and known as 'big John' because of his ample size, was invited by an uncle who owned a villa at Benedorm, to spend two weeks there in the summer, bringing as many friends as he liked. He took Nigel, Nick and Peter. This continued over several summers and on the beach the boys formed many friendships with numerous girls and young couples, all of whom turned up eventually at Court Hill. There were Bavarians Verne and his wife (whose name escapes me); Ev and Rhud who were Dutch; Polish Trisha, who brought us poppy bread and various delicacies; from Austria came studious Eva, who spoke five languages fluently and was an official guide in Vienna and her friend Lottie, a lively comedienne who enlivened us with her wit and laughter. Nick's return visit to her home must have been different from hers in our modest farmhouse, for her father was a diplomat who had an apartment in Schobrann Palace. Lottie's wedding, some years later, which Nick attended, was conducted in the palace chapel.

Lastly came Marlene whom Nick had met in her home town of Cologne. Her stay was memorable for one occasion.

"Today," she announced at breakfast time, "I cook the lunch, I make for you a typical German lunch, I will need your biggest saucepan."

Not averse to handing over the cooking, I looked out the saucepan in which I usually boiled the Christmas puddings. Then Nick took her to Arundel whence she returned with the largest cabbage she could find and a quart of vinegar. She also produced some very large German sausages, which she had brought with her.

I found her a chopping board and a large knife and left her slicing the cabbage. We were to be introduced to sauerkraut.

I had only seen it before in small jars in delicatessens. Marlene made enough to serve the crew of Nelson's flagship. Duggan, very conservative about food refused it politely, as did Jonathan. Peter took a teaspoonful. Nick, Marlene and I, partly out of politeness and partly curiosity, ample helpings.

Over several days the three of us, I having acquired a taste for it, finished the saucepanful. At least, I thought, we would not get scurvy. Marlene came again the next year, at Easter, with her sister Honey and on Easter Monday nearly all the other continental friends joined us for a picnic at Cowdray Point-to-Point Meeting.

"Can you do some ham and cold chicken?" Nick asked me a few days earlier. "But don't bring sandwiches."

Sandwiches, formerly my idea of a picnic with a thermos of coffee and an apple, were taboo. I cooked a gammon and a large chicken and all the others made a contribution, the day was a gourmet and social occasion. Only Dug and I were seriously interested in the racing. Marlene came several times over the next few years and I occasionally saw the others but Nick kept in touch with them; most of them came to his wedding.

The girl I remember with the most affection was Hermione. After a year in Oxfordshire, Nigel secured the post of assistant manager at The National Agricultural Centre at Stoneleigh, Warwickshire. Very soon he met Hermione, who lived in nearby Kenilworth.

One evening he telephoned to say that they were off on a three week tour of Portugal and wanted to spend the night at Court Hill before taking the ferry the next morning. Accordingly, late on Friday night a week later a new MGB drew up on the drive and there alighted from the driving seat a six-foot blonde accompanied by Nigel. In the following weeks she drove him all over Portugal for no one else was allowed in the driving seat or indeed to touch her MGB. She had taken a course in engineering in order to dismantle it and service it herself.

After this she often came to stay, quickly endearing herself to us with her outgoing personality, her sense of humour and love of the country. Once she came, not with Nigel but with Ramses, a huge, black, shaggy dog who, in keeping with his royal name, brought his own purple-cushioned bed and purple towel. Ramses was as exuberant as he was large and unused to country life, chased everything that moved; the hens until they took refuge in the hen house, even an indignant cockerel. Jess regarded him with tolerant amazement, then, on his last day, like some mischievous country child leading a naïve town cousin into trouble, she contrived to walk him into the sludge pit, without getting so much as a toe in it herself and brought him home with his thick coat matted with stinking, glutinous manure. We hosed him down in the yard but I do not think the purple towel ever looked the same again and the stench still clung to him when he left the next day.

For some reason Hermione was not driving the MGB but a borrowed Mini. As I watched her fold her long length into the driving seat with Ramses overflowing the seat beside her, his legs wrapped round the gear stick, I hoped that they would arrive home safely and that the car's owner would not object to a lingering smell.

Letters from Hermione told of Ramses further escapades until one day he leapt through the service hatch, taking most of a valuable dinner service with him. Her long-suffering parents finally put their combined feet down. Ramses must go and he went.

By now I had secretly nurtured hopes of Hermione as a daughter-in-law, but it did not happen. In time she married and went to live in America while Nigel became engaged to Jo – a tall, dark-haired girl with a long, pale face and often melancholy expression; she never seemed to fit in happily at Court Hill as so many other girls had done. The engagement did not last, neither did Hermione's marriage, but she settled down in New York where she became a stockbroker. She still writes to me today, affectionate, amusing and descriptive letters, thirty years later.

1972 saw the first wedding in the family. A few years before Peter and Mike had brought two girls to tea, Fiona who was partnered with Mike and Angela with Peter. After a few weeks they changed partners and Fiona became a regular addition to our Sunday teatime circle, which often numbered a dozen or more. Slim and attractive, with chestnut brown hair and large blue-grey eyes she was the second of the four daughters of an Army Colonel's widow who lived near Bognor. In due course we met the three sisters, Christine, their mother and their paternal grandmother. A year later Peter and Fiona became engaged. We did not attempt to dissuade them, to say that they were too young. Peter had always made his own decisions, which invariably turned out to be right for him. Neither did we need to offer advice when he bought a flat outside Worthing, obtaining a bridging loan from his bank and he and

Fiona fixed the wedding date for the last day in September. He was twenty-one, she was a year younger.

Nigel and Nick rallied round 'the guinea pig' as they called the first to launch into matrimony, with generous presents and any help they could offer in setting up the home.

Sponsored by a wealthy uncle, Fiona's wedding lacked nothing and bore no resemblance to the modest wartime marriage of Dug and myself, in a tiny country church with only my mother, the best man and two witnesses present. In 1944 austerity was at its peak, not even one bottle of wine was to be had and our few guests were provided with just coffee and sandwiches.

The only similarity was that both were on a perfect September day when sun shone warmly from a sky of cloudless blue. At the Royal Norfolk Hotel, facing a sparkling sea, champagne flowed freely, food was abundant and the three-tier cake was of model of confection.

But to go back to the early morning of that day, we were all up with the lark for the ceremony was at 12 noon and break-fasted in an atmosphere of suppressed excitement. Jess, aware as dogs quickly are, of anything unusual going on, retired to her basket and eyed us apprehensively over the rim. The boys went upstairs to help each other into the suits, which had been collected from Moss Bros. the day before and hung up overnight.

Peter came down, looking as nervous as Jess, the first time I had ever seen him not completely confident and debonair.

"Can you help me, Mum?" he asked, fumbling to get a stud into a stiff collar.

I did not find the operation easy but at last it was done.

"Can you keep a secret?"

"I always have."

"I've got to leave the address of the hotel where we are staying."

Because the pair had spent two weeks holiday in Tenerife the previous summer, the honeymoon was to be a modest affair of a few days in the New Forest.

"It wouldn't be the Balmer Lawn by any chance?" I enquired.

"How do you know?" he exclaimed in surprise.

"I didn't, I only thought it would be an odd coincidence if it were. As a Land Girl I used to deliver a ten gallon churn of milk at the kitchens every day back in 1941. It was an officers training college then."

Nigel, always last to be ready on any occasion, now joined his brother. I almost asked 'Have you got the ring' as he went out of the door, but decided that it was only in sit-coms that the best man forgets the ring.

Dug and I were just about to leave with my mother and Jonathan, enjoying a day home from boarding school to attend his brother's wedding, when Big John drove up.

"Nigel went without the ring," he announced. "It's on his chest-of-drawers."

I ran upstairs to fetch it while he turned the car. I passed it into his waiting hand and he tore away with spinning wheels and shower of gravel.

Because Fiona had been brought up in the Roman Catholic Church, the wedding ceremony was held in the Church of our Lady of Sorrows in Bognor Regis, conducted by the incumbent priest assisted by our Rector.

Fiona, arriving only a few minutes late on the arm of her uncle, looked lovely in a simple but elegant dress of white lace over silk, her hair and veil held back with a caplet of tiny flowers. She was attended by her two younger sisters, Georgina and Mandy (a slender fourteen-year-old whom Nick married ten years later) and a tiny bridesmaid and page, all dressed in shell pink.

There was one more mishap with the ring when the priest dropped it into the receptacle for the Holy Water, but at last it was safely on Fiona's finger.

Then came the taking of photographs outside the church until the bride was tucked gently into the waiting car by a smiling chauffer, who years before, as a taxi driver, had driven her daily to school. The reception went on for hours as receptions do, from the greeting of relatives, old friends and strangers at the receiving line to the send-off of the young couple, Peter somewhat perturbed by Mike and Nick's artistic handiwork on the new sports car lent to him by his firm for the honeymoon. Then came the final 'goodbyes'.

Because it was necessary for Fiona to continue her job at a Worthing bank, it was five years before we welcomed our first grandchild. Then another baby boy was brought to Court Hill. His name was Russell.

19.   *Friends and Dogs*

Against the odds, complicated by Dug spending several weeks in bed with a back injury – during which time I liaised between him and the farm staff – I finished the history; but the work was not yet over.

There were people to see who still had interesting things to tell me and documents I had failed to consult until I felt like saying with Huckleberry Finn.

> *"There ain't nothing more to write about,*
> *and I'm rotten glad of it, because if I'd knowed*
> *what a trouble it was to make a book*
> *I wouldn't have tackled it."*

Then there was the round of local printers looking for the best and most economical way of producing it, then the proof reading and lastly the marketing, another round, this time of booksellers some of whom were very cautious or a little disdainful of small books that would not have a big sale.

I looked in despair at my stock of a thousand books, in neat brown packages of twenty-five, stacked on the wide shelves in the larder where once the boys had carried on their hobbies. How would I ever sell them?

"They'll go," said Dug cheerfully and go they did. They had come out just before Christmas and solved many Christmas present problems.

The local Post Office did well for me while Dug was an excellent salesman. While I was reticent about mentioning my publication, Dug had no qualms about asking, "Have you read my wife's book?" He kept a few copies in the Land Rover to sell to walkers he met on the Downs. Early one Sunday morning, while going round the cattle, he met an Australian visitor whose ancestors had been born in Slindon and brought him home to breakfast. He bought several books to give to relatives in Australia.

It sold quite rapidly in the village and went fairly well in some bookshops.

Sales slowed down in time but by then I had broken even and looked forward to making small profit. That took a few years, but in 1974 I brought out a second edition. This was larger, containing a lot of information I had overlooked the first time.

One does not, however, make a fortune from writing local history. The real remuneration was the interest it brought into our lives, the people from all walks of life that we met and even lasting friendships that we made.

One of these was with Cherry and Bryan. One morning a young woman telephoned me. She was taking her final exams in photography at college and was doing a photographic thesis of a village. She had chosen Slindon and would I advise her on what to photograph. That afternoon she arrived, lively, attractive, titian hair to her waist, driving her father's sports car. We spread old maps and photographs and my book on the sitting room floor and got to work. Cherry became one of our many regular visitors to the farm, accompanied by a small, pretty but unruly and somewhat aggressive Welsh Springer bitch called Hannah. Nick showed an immediate interest but Cherry was soon to become engaged to a photojournalist Bryan Alexander. Soon afterward they married and worked the Arctic together. They visited us occasionally, with Hannah, but she did not become friendly, being excessively protective of her owners and any of their possessions, especially the car, not allowing anyone within yards of the vehicle. While admitting that she had a certain charm, we viewed Hannah with some trepidation.

We had been without a dog for some while. Jess at the age of six had developed a lump on her stomach, which proved to be malignant. The vet operated and she quickly recovered and on his advice, we mated her and she produced a fine litter of puppies. A year later another lump appeared on her left foreleg. The cancer had returned.

A long course of treatment had no effect, she became painfully lame and there was no alternative but to have her put down.

I did not feel like starting again with another puppy and Jon, although he had never known home without dogs would only miss them in the holidays.

Then in September that year friends asked if we would like a sixteen-month-old Golden Retriever bitch. They had relatives who were going aboard and wanted a home for her.

So Kerry came to us a few weeks later. She was very large for a bitch and pale in colour. Although of an affectionate temperament with a desire to please, she was madly exuberant and utterly unpredictable.

"She's untrainable," declared Dug who had been very reluctant to have her.

"No dog's untrainable," I protested with more optimism than I felt.

"Well, she's your concern. Just don't let her cause me any worry".

I had been warned, so I started straight away on the training. I took her for daily long walks on the lead, because she did not come when called and chased everything that moved, especially pheasants.

I walked her through the village, keeping her between my legs and the flint walls, tapping her smartly on the nose when it protruded past my knee. This said the experts, would cure her of pulling. In spite of this enforced discipline she went crazy with delight when I picked up her lead, leaping on the kitchen table and snatching off my headscarf. In time most of my scarves were minus a corner or two. This was only one of her mad antics, which inspired me to write my first attempt at blank verse, which I unashamedly reproduce here.

20. Kerry

She was sixteen months old when we had her,
Our fifth Golden Retriever, still a boisterous, overgrown pup,
Slobbering, fawning, leaping, running off with socks and
     gloves;
Not chewing them but leaving them wet and slimy and often to
     be searched for.
She must greet all callers with something in her mouth, her
     blanket or a towel or perhaps a duster would do.
But if, when these she cannot find, with great distress and
     frantic search
She sees the daily paper, neatly folded, as yet unread.
She seizes it and rushes to say 'Hello' shaking it, crumpling it,
     until it is taken from her, quite unreadable.
She is seldom clean. Wash off the mud, groom her until her
     coat is spun gold, then out again,
She wallows in the deepest slush and mire, splashes through
     muddy water, rolls in the muck returning happy and bedrag-
     gled as a cur.
She retrieves well, isn't gun shy,
Stays close when off the lead. She has her good points.
But such a puppy still, a full year later,
Chewing a stick to fragments on the carpet. I keep some neatly
     stacked
By the stove for kindling and she selects one,

Not near the top, but three parts down and tugging it out
Leaves the others scattered about the floor.
I guarded carefully my feather duster, of real ostrich plume,
    brought me from South Africa.
Then one day, to answer the phone, I put it down. She had it in
    a trice. The door was open.
She was away with it across the field, tossing it, shaking it,
    dragging it in the mud.
I followed helplessly calling "Kerry" "Dead" quite uselessly.
She gambolled on with her prize, until she reached the fringe of
    the wood.
There she spied other feathers on the wing and gave chase.
Left my duster lying there for me to fetch.
What a dog! Will she never learn? She looks up,
All drooping ears and soft brown eyes.
Oh well, perhaps there's still time yet…

In time, she learned to walk to heel; she would 'sit' and 'stay' and come to the silent whistle. I was winning.

Then we heard from Cherry and Bryan. They were going to Greenland for nine months to live with and photograph Eskimos. Would we look after Hannah?

To my surprise Dug agreed and one day early in January, Hannah arrived and we were soon to learn just how much one small, determined Springer bitch could disrupt our lives.

Kerry was delighted. She now had a partner in crime, even a ringleader, and all my careful training went out of the window.

Hannah soon became as possessive of us and ours as she had been of Cherry and Bryan and barked long and furiously at anyone who came near the house until even the farm staff were nervous of coming to the door. Our visiting family were not exempt and while it was good to have a housedog we did not wish to be protected from our nearest and dearest.

Hannah brought her own smart basket and blanket. Surprisingly she was not possessive about this, allowing Kerry to try it for size and comfort while she reclined in Kerry's. They exchanged frequently and never seemed to decide which they liked best.

She proved to be an escapologist extraordinaire. Slipping out when a door was opened the merest degree, she would be away across the fields and hunting in the woods so that I had to drop whatever I was doing and give chase. This often happened late in the day and many a dark, wet, windy evening was spent combing the woods for Hannah. Jon who had now finished prep school and attended a day school in Chichester was a great support on these occasions.

This wily little bitch was as clever at breaking in as she was at breaking out. She found a small hole in the chicken run, which even the hens had not discovered and at first we could not trace. Consequently the hens lived dangerously whenever they emerged from the house. Fortunately they escaped with their lives if minus a few tail feathers.

She and Kerry played endlessly together and I derived much amusement watching them romp on the lawn. After a chase the length of the walled garden they would bound into the kennel, a disused aviary, with a force that made the whole structure rock. In spite of this, in the spring a wren nested in a top corner, sitting tight throughout the entire disturbance and hatched a brood.

In the afternoons I walked them, usually on the lead, through the woods or over the Downs.

Hannah was another puller and with Kerry, her walking to heel forgotten, they pulled like a team of express ponies while I streamed along behind, returning home breathless and exhausted. But in time Hannah became one of the family and her behaviour improved.

In those early days when we were still strangers I would never have believed that in a few months she would sit contentedly beside me while I worked in the garden without a thought of running off. One thing I learned about Springer's, however difficult they may be to train or control once you have built up an empathy with them you can do anything with them.

There were still the odd occasions when she could not resist the field of barley by the house where she was completely concealed and I could only trace her whereabouts by the movement of the corn, or now and then a pair of brown ears, flung sideways like plover's wings above the golden awns.

Meanwhile letters from Cherry brought a new interest into our lives with a graphic picture of life with the Eskimos and the wildlife of Greenland.

February: Temperature 30c after warmer period when it had actually rained. One aspect of the warmer weather was the shortage of food, the worst effected being the huskies, normally only fed twice a week at that time of year. In spite of the food shortage they had been invited for a meal by an elderly couple Bryan knew from a previous visit, so that Cherry could try seal meat.

March: North for the walrus and while Bryan was away hunting Cherry stayed with an Eskimo family and brushed up on the language. In the big city they had lived in an insulated Nissan hut without water or loo and where Cherry baked bread in a saucepan over a paraffin stove. While walrus hunting Bryan and an Eskimo spent three days floating around on an ice floe, which had broken away from the main ice, until rescued by helicopter.

While in the North they lived in a primitive house eating mostly local produce, walrus, seal, Polar bear, Arctic hare and halibut all of which they found very tasty.

May: The arrival of spring with the first pair of snow buntings. Within a week came the little auks. Soon the sky would be black with them. The Eskimos found them tasty eating.

June: Twenty-four hours of daylight and the start of the thaw. By the end of the month the sea ice would be finished and until August or later there is open water; time to fish for salmon and hunt narwhales.

July: Most of the snow had gone from the land but there was still ice on the sea. Cherry had been working alone at Siorapalut. A trip into town, taking six to eight hours when the ice was good, had taken twenty. Shortly it would be impossible to make the trip by dog sled and Siorapalut would remain isolated until the ice went and boats could be used. Bryan was working at the end of the fjord, photographing nesting Snow Geese, while Cherry travelled North with a hunter and his family and two crew to collect eggs and down from Eider Ducks and hunt sleeping walrus; two to three hundred asleep on the drift ice. They had booked a flight to Copenhagen on 6th September and hoped to be home soon afterward. They were very sad to leave the Eskimos who were so friendly.

In mid-September they were back to an ecstatic welcome from Hannah who squirmed and grovelled, almost tying herself in knots in her delight. Her people had come back!

She had no further use for me. When I went to the open car window to say goodbye, she flew at it barking aggressively. No matter my loving care of her for best part of a year when she had sometimes curled up beside me on my bed, she was again with her people and she must protect them even from me.

Life seemed quiet without Hannah, in spite of the boisterous Kerry, who seemed not to miss her too much, perhaps glad to have the place and all the attention to herself.

21. *Florida*

In 1971 Dug's sister, her son and his wife came from Lakeland, Florida, to stay for two weeks and the following year we
made a return visit.

On 24th May we boarded a Boeing 707 at Heathrow. It was
my first long flight and in those days took a lot longer than it
would today – even longer because of a diversion. Soon after
take off the Captain announced a change of course owing to a
cyclone over the Atlantic. We flew across Pembrokeshire's
lovely countryside, over Ireland with its tiny fields looking like
a patchwork quilt from the air and miles of green countryside.
Not for nothing, I thought, is it called 'The Emerald Isle'.

Approaching the American continent we flew first along
the coast of Labrador then over Boston and New York, flying
low over the Statue of Liberty, Manhattan and New Jersey all
of which was so clear from 35,000 feet even the cars, then
Long Island with its beautiful beaches and holiday homes and
finally landed at Miami.

As Dug and I were like Babes in the Wood when it came to
travel, Peter and Fiona had taken us to Heathrow and put us
on the plane, but at Miami we were on our own. We had two
hours to wait for our connection to Tampa. We were flying
BAOC, which Dug insisted on erroneously calling BOCM
(British Oil and Cake Manufacturers) to the complete bewilderment of all officials.

Aimlessly we wandered over to the Pan American Airlines desk. The man behind smiled.

"Will you wait for the BOAC connection or will we fly with us? A plane is leaving for Tampa in half-an-hour."

Half an hour, we thought, was better than a two hour wait.

"We'll fly with you," we said.

It was not until we were on the plane that we remembered being told "On no account change airlines at Miami".

It was a very small plane that we boarded and the passengers looked like country people. One stout woman in voluminous straw hat boasting a wealth of fruit and flowers round the brim clambered on with a loaded shopping basket, others had bags and baskets. There were few suitcases beside our own.

The little plane juddered into the air and throughout the flight touched down at every little airstrip, where people got on and off. It was like a village bus. Eventually it landed at Tampa, then the newest and finest airport in the world, but at a different bay from the one where our relatives were waiting. It was only after considerable delay and much paging over the intercom, that we were united.

There was still a forty-mile car drive to Lakeland and it was dark by the time we reached the complex of flats into which Tet had moved two weeks before from Pennsylvania to live near to her son.

We found our itinerary for the next three weeks pinned to our wardrobe cupboard door. We read it and gasped. It looked as demanding as a Queen's tour. The following morning we were allowed to rest. We swam and relaxed by the large swimming pool in the centre of the complex. As Tet was as yet the only person to move in we had the pool to ourselves.

The janitor came out and tested the temperature of the water. "Only 72 degrees," he told us. "Kinda cold this morning."

I did not tell him I had swum quite happily in a water temperature of 52 degrees.

In the evening we were taken to have dinner with Clifford and Paula at their spacious bungalow home and met their two teenage daughters, Wendy and Cindy and ten-year-old son Rees. Rees had been fishing in the lake at the end of the toad and proudly displayed a small turtle he had caught.

Dug went out with Clifford the next morning leaving me with Tet. The extreme humidity of the Florida climate, like permanent Turkish bath, had a disastrous effect on my hair, which normally had a natural wave and was fairly manageable. My sister-in-law took one horrified glance at its limp strands and marched me off to the hairdressers where American women seem to spend a disproportionate amount of time, always off to the hairdresser looking as if they had just come from there.

"Your hair is that fine," remarked the girl who had washed it and left me unhappily sitting under a hot drier in a row of other clients for what seemed like hours. "I'll have to tease it good."

Tease it she did and sprayed it, so that I came away feeling as if I was wearing a bearskin helmet. The result of all this effort lasted all of 24 hours, but at least my appearance satisfied Tet at the open house party organised for us by Clifford and Paula at the yacht club that evening. We were to meet some of their friends – who included many notable medical men and their wives and Bob Wiley, Colonel of the Strategic Air Defence and a prominent businessman, whom I liked enormously. I was to meet him again thirty years later North Carolina.

In spite of our intensive programme, we managed most days to spend a little time relaxing by the pool.

"You don't come three thousand miles to sit by a swimming pool," sternly admonished Tet. But at home we did not have one outside the front door. Some evenings we spent playing pool with the girls at their home.

Clifford or Paula drove us on long excursions but for shorter journeys Tet drove us, which we found quite an experience for Tet made up her own rules of the road. She would stop at a traffic light to go to a shop impervious to the furious hooting from drivers behind. After dark she drove on the central white line. Lakeland residents had learned to keep out of the way of her big, black saloon. As frightening were the carloads of teenagers. Driving was permitted at the age of fourteen. Frequently we saw Wendy, driving her father's Ford Mustang, waving to us out of the window as she took a bend at speed, companions piled round her.

The highlight of our stay was the visit to Daytona Beach and the Kennedy Space Centre.

On arriving at Daytona we booked in at the Hawaiian Inn and made for the beach, where Clifford hired two Honda motorbikes. With Paula and me on the pillions, he and Dug raced twelve miles on sands that were so hard that motorcar races were held on them.

We swam in the sea and the hotel pool and watched the flocks of brown pelicans fly up and down in formation over the waves.

"We call them the Daytona Air Force," Paula told us as we looked for golden sand lizards.

After dinner in Clifford's suite we went on to a Hawaiian nightclub where we danced and watched Polynesian belly dancers. It all seemed very much removed from life on the farm.

The following morning we drove to Cape Kennedy. Clifford had injured his foot in the sea before breakfast and was in considerable pain, which he bore stoically throughout the long day. It began with a two-and-a-half hour bus tour of the centre.

Starting at the Tourist Information Centre we saw the original Apollo 7, the first manned capsule, blackened and scarred from its re-entry into earth's atmosphere and ending at the vehicle assembly building, 525 feet high and covering eight acres where the Apollo/Saturn exploration vehicles were prepared for launch.

Dug thought it would make a splendid grain store.

Since those days science and technology and space travel has made enormous strides but even then it was tremendously impressive.

"But what benefit is all this to our lives?" Dug queried.

"Well, to date," said Clifford, "The spin-off of space technology has brought advances in the fields of industry, medicine, meteorology, navigation of ships and aircraft and in agriculture. The possible application of satellite observation can help in crop disease and insect detection, irrigation development and many other aspects of farming and forestry."

We considered our question answered.

The acres of swamp surrounding the centre were a haven for wildlife. Two hundred species of birds inhabited them and many alligators. Paula drove the two hundred mile journey home and having dropped the children and us took Clifford to hospital where an x-ray showed the foot to be broken. He spent the next few weeks on crutches with his foot and leg in plaster. Due to this misfortune, two of our major trips had to be cancelled. One was tuna fishing from Clifford's yacht in the Gulf of Mexico, the other a trip to the Everglades.

But we did get to the Gulf of Mexico when Paula drove the family and us to Clearwater to see her parents. When we arrived, the girls suggested a swim in the sea before lunch, the sea being forty miles away. On reaching at the beach we witnessed a violent electrical storm on the horizon, constant vivid forked lightening played against a wall of purple-black sky. Lifeguards were keeping would-be bathers out of the water, but it soon cleared and we swam in a sea that was anything but clear water. It was the colour and consistency of pea soup and so buoyant with salt that it was impossible to dive or swim underwater. The beach was not the golden sand and palm trees of the picture postcards but dirty and lined with trash cans overflowing with popcorn packets. Along the edge, however, Rees and I collected some exquisite tiny shells of many shapes and colours, then back to the house and a shower to wash off the salt before the meal.

Staying with relatives gave us the opportunity for a more intimate look into life of this youngest of American states than has the casual tourist.

We were entertained in several homes, attended meetings and cookouts of the local cubs and scouts and lunched with children at a primary school. Because Dug was a governor of our local primary at home, he was invited to talk to a class about education in England, while I sat in at a science lesson.

We went to a citrus auction and an orange juice plant for Lakeland was the centre of the citrus belt, which extended almost from coast to coast.

The estates of new homes were built on former orange groves, so that most residents had oranges and grapefruit growing in their gardens. These are left to fall off the tree and rot, or for the squirrels, which gouge out the pips as cleanly as a knife and leave the rest, while the families orange juice comes from tins of frozen concentrate.

The waste peel and pulp from the manufacture of orange juice is fed to cattle. We saw this in its raw state put out in the fields at a ranch of Brahmas, the beef cattle of Florida. For the milking herds of Friesians and Jerseys it was used in a dried form mixed with other feedstuff.

Married women lived a very full life mainly centred round their homes and families. They entertained a great deal; inviting friends over for dessert after the evening meal was one of the simpler ways and a novel one to us. Because of the long distances they spent many hours a day driving to the big 'park and shop' centres, taking their children to school, to riding, swimming and various classes. They were intensely involved in all their children's activities. Whether or not their young son's team won at junior baseball game can make their day.

We were taken to see Rees play in the Junior League Baseball and to a professional baseball game, where Paula, beside me, explained every move and rule until I knew more about American baseball than I did about British Soccer.

The older women gave numerous tea parties at their homes for their friends, one of whom was invited to pour the tea. The next week a piece always appeared in the social column of the local paper, as follows:

'Mrs Morgan gave a tea party at her home' it then listed those present, 'Mrs So and so poured'. Since then, if I am ever asked to pour the tea in someone else's house I always think 'and Mrs Duggan Rees poured'.

A piece appeared in this column about our visit. It said that I was affiliated to the National Trust and had published a book on its activities, which was not entirely accurate!

Our last visit was to the Cypress gardens, which we toured mainly by boat. The flowerbeds were filled with colourful, exotic plants and pretty girls in pink or blue crinoline dresses stood waiting to be photographed with tourists. Cypress trees grew out of the water, which was inhabited by the ubiquitous alligator. These were found even in the lakes in the town of Lakeland. I thought them unprepossessing creatures especially after one crawled out of the water and devoured an old lady's pet dog.

One evening we went for drinks with a millionaire and his wife. Dollar millionaires were commonplace in Lakeland. In their sumptuous home, on the elegant, marble stairway that reminded me of something out of a Broadway movie show, I encountered a large, shiny cockroach, which invaded even the most immaculate homes.

The couple had recently been burgled and this was still a topic of conversation.

"These two men tied us up in our chairs," our hostess told us, "while they ransacked the house."

"I jest must go to the toilet," I told one of them, "so he picked me up and carried me, trussed up as I was and sat me on the toilet!"

The police rescued them sometime later after a friend who called raised the alarm.

We visited restaurants where the steaks were as big as dinner plates; went to church in a new building, all pinewood and thick, blue carpet like many others all over Lakeland bequeathed by wealthy citizens.

The flight home was not without incident. We negotiated the change of plane at Miami and boarded the plane for Heathrow. We had barely taken our seats when the Captain announced that we must disembark as he had received a report of a bomb on board. We did so with alacrity and were led to a lounge where we were all frisked and questioned then waited two hours before, weary eyed we were taken back to the plane. The bomb scare, we were told, was a hoax by some drunken Argentineans whom we saw being marched off by police. We prepared to board but again came the announcement that two passengers refused to fly and we must wait while their luggage was taken out. Finally, the Captain persuaded them amid cheers and applause and at long last we were airborne to be greeted at Heathrow by our anxious relatives.

22. *The End of an Era*

Six years passed and we began to think of Dug's retirement. He would be sixty-five in the autumn and would have completed twenty-seven years of hard work managing the farm and shoot.

With a move in sight I began, in what time I could spare from the general routine, to clear out the four rooms and large cupboards at the top of the house. Since this floor had no longer been used everything that was not needed had been taken 'up to the top'. The result was a conglomeration of old sports kit, shoes, jackets, books and pictures. I gazed round and wondered where to start.

First the tank room with its capacious cupboards full of tea chests, boxes and old cases. These contained all the memorabilia of childhood, toys, nursery pictures, photographs and the occasional skeleton of a rat that had infested the cupboards when we first came. Every box had its compliment of Lego. From everything I took out fell a shower of little, white, plastic pieces. I washed them and put then in a box for Russell, likewise well-distributed sections of rail and items of a Hornsby train set. There was two pairs of small boxing gloves, somewhat tattered, handed down three generations that the boys had sparred with years before and a collection of soft toys, which I washed and mended, the best to give to Russell, others to a jumble sale. Among them was my old teddy bear, which I had at the age of one year. He had lost his eyes and much of his hair but retained his full compliment of limbs and ears. In another chest was the knitted suit that had been his daywear and the striped pyjamas he had worn at night.

What to do with him? I could not give him to the grand-children who would not treat him with the respect to which he had been accustomed and in our future domain our lim-ited cupboard space would be full.

After some thought, I decided to cremate him. Having built a funeral pyre in the garden I wrapped him in a piece of sheet, laid him gently on top and set a match. Accompanied by Kerry and two cats I watched it burn; the last relic of my childhood.

Later, on Antiques Road Show I saw a similar bear, in not much better condition, valued at quite a large sum. Oh well, we all make mistakes.

It was difficult while turning out not to indulge in trips down memory lane and to decide what to keep and what to discard.

"You've got to be ruthless Mother," Nigel told me, on a visit home. "What any of us haven't seen for about twenty years, we are not likely to miss."

Such was not the case.

"Oh, you haven't thrown that away!" was often the cry.

One thing I did inadvertently throw out from an overflow-ing drawer in the big desk in one bedroom was a bundle of Nigel's letters from Norway and Denmark, This I regretted.

By the end of the year friends were inquiring and we our-selves were wondering, where we were going to live when we left the farmhouse. We wanted to stay in Slindon, but properties in the village were prohibitively expensive. The National Trust offered us a pleasant house in Top Road with a view overlooking the village to the sea, which we liked, but a new agent put up the rent to an unaffordable amount. I alone was not worried. With Macawberish optimism I maintained that the right thing would turn up when the time came ... and it did. Many years before, we had met two retired nurses who lived in a small bungalow on Slindon Common, the other side of the main road from the village. We had become friends and often visited them at Ray Cottage. We fell in love with the little place, cosy, compact and with a certain character about it and set in two thirds of an acre of beautiful garden.

"If you ever want to sell this place," I said, half jokingly, "Will you give us first refusal?" – never thinking they would ever leave the little nest of which they were so fond.

Molly was a writer, the gardener and driver while Eileen shopped in the village on foot, cooked and did all the household chores. Quite suddenly her sight began to fail and she developed a heart condition. She could no longer see to cross the busy road or climb the hill to the village. Regretfully they must move into the town to be near shops and on level ground.

Molly was as happy for us to have the cottage as we were to move there, knowing she said that it and especially the garden would be in good and loving hands. Furthermore, Mr Long agreed to buy the property in which we would live, rent free, for Duggan's lifetime.

Molly and Eileen moved out in July and the cottage, as we always referred to the bungalow because of its 'cottagey' feel, would be redecorated for us to move in at the end of September. Then "Mr Long wants me to stay for another six months". Dug said at breakfast one morning toward the end of this time.

"Certainly not," I replied. "I won't let you."

He had suffered a lot of illness during the past few years, and was more than ready for retirement. But I was over-ruled. It was agreed that Dug should carry on until April. We faced another winter in the cold farmhouse, while our cosy cottage remained empty.

During the remaining summer months and the autumn I drove down to the cottage frequently to work in the garden, discovering new treasures planted by Molly's green fingers and to pick the lush blackberries that grew in the hedge surrounding it, for years before the plot had been part of a common. Kerry came with me and explored and approved her new territory.

A hard winter followed. Heavy snow fell at the end of January and lasted into February in freezing conditions. One morning I found water in the airing cupboard and more pouring down beside the Rayburn from burst pipes.

There were brighter moments. Marlene came to stay and Nick came home for a long weekend while she was with us. Cherry visited us with Hannah and her daughter Pippa.

During this time we gradually sold all our big furniture, Dug with sadness as much of it had come from his old home, but there was no way it would fit into the smaller rooms of our new abode. It was fun, though, going round antique shops again choosing more suitable pieces. We moved at the end of February, two months before Dug finished work. As the tail-board of the single removal lorry slammed to and the vehicle lumbered down the drive and out of the white gates, I glanced back at the house.

It looked forlorn, bereft of life, its un-curtained windows staring and blank. Did a house have a personality of its own, I wondered, or did it merely take on the personality of its inhabitants – of our family and the families that had gone before us – Hilaire Belloc's and others? Would anything remain of us in this old house, of our boys or of all those other young people who had joined them over the years? I hoped so and that it would continue to be a happy house.

"Come on Jo..." called Dug from the waiting car.

End

More books of local interest from Woodfield Publishing

A Portrait of Slindon | Josephine Duggan Rees

Recounting lively stories of Slindon's former inhabitants over the centuries and of its many interesting and historic connections, this book contains a wealth of information and records for posterity the amazing richness and variety of village life in this little Sussex village over many centuries. This is a completely revised edition of Josephine Duggan Rees's popular book, now professionally published for the first time and containing many additional photographs and updated text.

ISBN 1 903953 57 8 | 236 softback £12.00

Corduroy Days by Josephine Duggan Rees

The author recalls her younger days as a teenage volunteer in the Women's Land Army during World War II in a succession of warm-hearted and gently humorous stories which follow her progress from clueless city-girl to proficient countrywoman, on the way meeting up with a host of colourful rural characters and Duggan, her husband to be. Her delightful depiction of their less-than-romantic courtship is one of the book's many highlights.

ISBN 1 873203 48 9 | 226 pp with photos | softback £9.95

20th Century Farmer's Boy by Nick Adames

Nick Adames' unravels the fascinating history of his family, who have farmed in Sussex for over 400 years and still own farms at both Flansham and Madehurst. He entertainingly recounts their fortunes and misfortunes during the last century and recalls the many local people and events they have been involved with over the years. Older Slindonians will doubtless remember 'Buckle' Adames, Nick's uncle, a well-known local character, whose eccentric antics are among the many tales of local interest to be found in this enjoyable book.

ISBN 1-903953-01-4 | 388 pp with photos | softback £12.00

Just Visiting by Molly Corbally

Molly Corbally, who was a well-known resident of Slindon for many years, tells of her former life as a District Health Visitor in the early years of that profession, the 1950s & 60s, when the villages around Kenilworth in the Midlands comprised her 'district'. In a series of funny, warm and insightful stories, set in a post-war England now vanished, she describes the many weird and wonderful personalities her professional life bought her into contact with and how she dealt with their many eccentricities and predicaments.

ISBN 1-903953-06-5 | 300 pp | softback £9.95